Purnell's
Pictorial Encyclopedia

Edited by Theodore Rowland-Entwistle

Purnell

This magnificent carving
of a weeping figure forms
part of a totem pole in an
American Indian village
near Hazelton, British
Columbia.

Executive Editor
Campbell L. Goldsmid
Compiled and edited by
Theodore Rowland-Entwistle, B.A., F.R.G.S. and
Jean Cooke
With contributions from :
Neil Ardley, B.Sc.
Jill Bailey, M.Sc.
John Borwick, B.Sc.
Arthur Butterfield
Ronald S. Carter, B.Sc. (Econ), F.R.G.S.
Marshall Coombs, M.A.
Ann Kramer
Mark Lambert
Keith Lye, B.A., F.R.G.S.
Jonathan Ormond, B.A.
Keith Wicks

Picture Research by
Anne-Marie Ehrlich, B.A.
Layout
Fizz Robson
Lindsey Rhodes

SBN 361 04416 X
Copyright © 1979 Purnell and Sons Limited
Published 1979 by Purnell Books, Berkshire
House, Queen Street, Maidenhead, Berkshire
Made and printed in Great Britain by Purnell
and Sons Limited, Paulton (Bristol) and
London

Designed and produced for Purnell Books by
Autumn Publishing Limited, 10 Eastgate
Square, Chichester, Sussex

Contents

EARTH AND THE UNIVERSE 6
Earth's Surface: The Land 8
Earth's Surface: The Sea 10
Inside the Earth 12
The Moon, Earth's Neighbour 14
The Wandering Planets 16
The Life-giving Sun 18
The Distant Stars 20
Earth and Universe Inquiry Desk 22
Exploring Space 24

THE LIVING WORLD 26
The Plant Kingdom 28
The Animal Kingdom 30
Plants and Animals of the Oceans 32
Plants and Animals of Europe 34
Plants and Animals of Asia 36
Plants and Animals of North America 38
Plants and Animals of South America 40
Plants and Animals of Africa 42
Plants and Animals of Australia 44
Plants and Animals of Deserts 46
Plants and Animals of the Poles 48
Living World Inquiry Desk 50
The Human Body 52
The Descent of Man 54
Human Body Inquiry Desk 56
The Food Chain 58

THE WORLD OF IDEAS 60
The Great Religions 62
The Great Philosophers 64
Myths and Legends 66
Sharing Ideas 72
Producing a Book 74
Ideas Inquiry Desk 76
The Story of a Newspaper 78
Radio and Television 80

SOUNDS AND PICTURES 82
Making Music 84
Musicians at Work 86
Recording Sounds 88
Acting a Play 90
Drama Through the Ages 92
Taking Photographs 94
Making Movies 98

Film-Makers at Work 100
Arts Inquiry Desk 102
Making Pottery 104
Making Sculptures 106
Sculptors at Work 108
Making Pictures 110
Artists at Work 112
Gallery of Art 114

PEOPLE AND PLACES 116
People of Europe 118
People of Asia 120
People of North America 122
People of South America 124
People of Africa 126
People of Australasia 128
People of Oceania 129
Languages 130
People and Places Inquiry Desk 132
Highlights of History 134
History Inquiry Desk 138

SCIENCE AND TECHNOLOGY 140
Oil and Natural Gas 142
Generating Power 144
Coal from Ancient Swamps 146
Alternative Energy 148
Building a Home 150
Water—the Peculiar Liquid 152
Perils of Pollution 154
Soaps and Detergents 156
Science Inquiry Desk 1 158
Fun with Calculators 160
Using a Microscope 162
The Mighty Atom 164
The Story of Plastics 166
Lasers: Light at Work 168
Using Metals 170
Making Textiles 172
Energy and Waves 174
Science Inquiry Desk 2 176
Rolling Along the Road 178
Travelling by Train 180
Flying Through the Air 182
Hydrofoils and Hovercraft 184
Travelling Through Space 186

INDEX 188

Earth and the Universe

If you take a long journey—by road, rail, air, or sea—you soon discover that the Earth is a very big place. The more slowly you go, the bigger it seems, but large though it is, the Earth is very tiny indeed compared with the rest of the Universe, the mass of stars and planets that float in space.

It is very hard for anyone to grasp just how big the Universe is. Fortunately Earth has some comparatively near neighbours, and we can understand how far away they are much more easily. From that basis we can work out the distances of other bodies in outer space.

Earth's nearest neighbour is the Moon, about which you can read on pages 14-15. The

and round it in orbit. The Sun is a star, a heavenly body that gives off light. The planets and the Moon, although they too seem to be lights in the sky, do not make their own light. They reflect the light of the Sun, like so many huge mirrors in space.

The Sun, with its attendant planets, forms part of a huge array of stars which is called the Galaxy or the Milky Way. A group of this kind is called a star system. The Milky Way is disc-shaped. It contains many millions of stars, which revolve around the centre of the disc, just as the planets revolve around the Sun. There are many other star systems, also called galaxies, but only three of them can be seen with the naked eye.

Nobody knows how old the Universe is or how it began. However, people have many

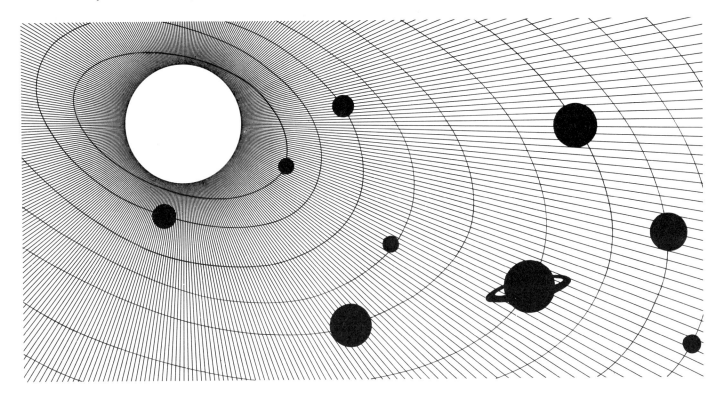

Moon is about 385,000 kilometres away. To give you some idea of how far that is, if you were to travel there as fast as you fly in a jumbo jet, it would take you 26 times as long as it would to fly from London to Sydney—right across the world from one side to the other.

The Moon is really very close to us. The next nearest body in space is the planet Venus, which is 41,360,000 kilometres distant when it is at its closest. To travel there at the speed of a jumbo-jet would take you about 2,800 times as long as it would to fly from London to Sydney. A journey to the Sun at the same speed would take you more than 10,000 times as long as the London-Sydney flight would.

Earth and its neighbours, the planets, are satellites of the Sun—that is, they circle round

The solar system. In this drawing the planets are shown in their relative sizes, but the Sun should be much bigger, and so should the distances between the planets. We would need a page as big as a football pitch to show them properly! In order from the Sun the planets are Mercury, Venus, Earth, Mars, Jupiter, Saturn, Uranus, Neptune and Pluto.

theories about it. The most popular is that all the matter of which the Universe was made was originally condensed into a lump about 30 kilometres across. This lump exploded with a 'Big Bang' about 15,000 million years ago. Scientists think this idea is probably right because the Universe is still growing bigger as stars and galaxies move further and further away from one another. The Earth itself is thought to be about 4,500 million years old.

What distinguishes the Earth from all other heavenly bodies about which we know anything is the presence of life on it. Life as we know it needs two things for its existence: an atmosphere containing oxygen, and water. Life began on Earth some time during the first thousand million years of its existence. We know this because we have found remains of living organisms which were fossilised (turned

to stone) about 3,500 million years ago. In the early days of the Earth there was very little oxygen in the Earth's atmosphere, but there was a lot of methane and other gases. However, a great deal of the Earth's surface was covered with water, just as it is today. Radiation from the Sun acted on chemicals in the water to make complex chemical compounds. Gradually these compounds joined together to form even more complex substances, until eventually living cells were formed.

In time, living organisms developed which could make their own food, as plants do today (see pages 26-27). In doing so they gave off oxygen, and this was the origin of most of the oxygen that is present in the Earth's atmosphere today.

Is there life in space? We do not know, but the answer to this question is almost certainly

The Earth from space—a magnificent photograph taken from the Apollo 17 spacecraft, which made the last trip to the Moon in December 1972. The picture extends from the Mediterranean Sea to Antarctica, which is shrouded by heavy cloud. You can see the coastlines of Africa and the Arabian Peninsula very clearly.

'yes'. For life as we know it the conditions must be just right: not too hot, not too cold, with some water. The planets nearer the Sun than we are, Mercury and Venus, are both too hot. Mars, the next planet out from the Sun after Earth, is the only one where life may exist. It has a very thin atmosphere, with a little oxygen, and it has certainly had water in the past. An American space probe which landed on Mars found no clear evidence of life, or proof that there was no life.

However, other stars may have planets as the Sun does, and some of these may well be like Earth and have life on them. In 1961 two meteorites—pieces of rock from outer space —were found showing what appeared to be the fossilised remains of very simple living things, so there does seem to be life of some kind outside the Earth.

Earth's Surface: The Land

The part of the Earth's surface most people know best is the land, although it covers only 30 per cent. of the world. It is made up of seven continents: Europe, Asia, Africa, North America, South America, Australia and Antarctica; and thousands of small islands. You can read about the differences between these continents, and about the people who live in them, on pages 116-129; and about their plant and animal life on pages 34-49.

All these continents, however, are made up of the same basic ingredients: mountains, valleys, rivers, lakes, deserts, plateaus and plains. Man has changed the face of the Earth considerably in the short time he has lived on it, but natural forces have made much bigger changes. These changes are still going on, but they take place so slowly that most people are not aware that they are happening.

The most violent changes to the Earth's surface are the result of forces inside the Earth, some of which we can see as volcanoes and earthquakes, as described on pages 12-13. Some changes happen comparatively quickly. For example, the volcanic island of Surtsey, off the coast of Iceland, was created in 18 months in 1963-65. Other changes are much slower. For example, the southern part of England is slowly sinking, but it will be many thousands of years before it is under the sea.

Frequently parts of the Earth's surface are lifted up as a result of underground activity in the Earth's crust. In this way mountains are formed. Sometimes the mountains are made by folding of the crust, in much the same way as a pile of blankets folds if you press lightly on each end. Sometimes a whole block of ground is lifted up by pressure from underneath. Other mountains are made by volcanic action. Surtsey is actually the top of a mountain which starts under the sea.

When they are new, mountains are rough and jagged. Over millions of years these rough features are worn away by rain and wind. Water running down a mountain gradually wears it smooth. The water also helps to break up the rocks. If very wet rock freezes the water in it expands and cracks the rock. In time rock breaks down into tiny particles which form soil. Soil is made richer by rotting plant and animal matter mixed into it. Rivers carry fine particles in their waters as they rush towards the sea. As they reach the sea they flow more slowly and deposit their soil particles to form sediment.

In time there comes another upheaval in the Earth's crust. The layers of sediment are crushed together to make new rocks, called sedimentary rocks. Sandstone is a good exam-

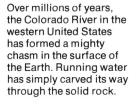

Over millions of years, the Colorado River in the western United States has formed a mighty chasm in the surface of the Earth. Running water has simply carved its way through the solid rock.

ple of such a rock. Eventually some of these sedimentary rocks are forced up from the seabed to form new mountains.

Buried in the sediment are the bodies of countless sea creatures. Most of the soft part of their bodies decays, but the hard parts—skeletons and shells—survive and turn into rock, as fossils. This is why you find the fossils of sea creatures in rocks on land.

The process by which rocks are worn away is known as erosion. Rain, rivers and the sea are not the only forces of erosion. The wind also plays a part. It can blow fine particles against a rock face to wear it away. In deserts you can see many rocks which have been cut

away by continually having sand blown against them. Glaciers, which are rivers of ice, also wear away rock. They move very slowly but they can cut away the surface of a mountain in much the same way as a carpenter's plane smooths wood.

We have seen how sedimentary rocks are formed, but there are two other kinds of rocks. Igneous rocks get their name from a Latin word meaning fire, because they are formed inside the Earth under great heat which melts them. Granite is an example of igneous rock. The third kind of rock is metamorphic rock, which gets its name from a Greek word meaning changing form.

White Island lies off the northern coast of New Zealand's North Island. It is an active volcano. Water from geysers runs into its crater, sending up a column of steam.

Metamorphic rock has been reprocessed by the action of heat and pressure inside the Earth. For example, marble is a metamorphic rock formed from limestone, a kind of sedimentary rock.

Plants also help to reshape the surface of the Earth. The root of a tree can find its way into a tiny crack in a rock. As it grows it acts like a wedge, splitting the rock apart. Smaller plants such as grasses help to cover the soil and prevent the wind and rain from removing it, while trees and bigger plants help to protect the grasses. If all the plants are removed the wind dries all the moisture out of the topsoil and blows it away as dust, leaving a desert.

9

Earth's Surface: The Sea

More than 70 per cent. of the Earth's surface is covered by the great oceans and the lesser seas that are linked to them. There are four oceans: the Pacific, Atlantic, Indian and Arctic Oceans. Some people describe the southern parts of the Pacific, Atlantic and Indian Oceans as the Antarctic Ocean. Around the edges of the oceans there are also several seas, most of which are surrounded partly by land. Examples are the Caribbean Sea, the Mediterranean Sea and the China Sea.

The depth of the ocean varies a great deal. Near land the sea floor is the continental shelf—part of the land which is under water but is attached to the continents. Waters over the continental shelf are comparatively shallow, but those over the main ocean bed are often very deep. The average depth of the Atlantic and Pacific oceans is about 4,200 metres. The Indian Ocean is not quite so deep, while the Arctic Ocean has an average depth of only about 1,500 metres. Some parts of

Right: A whale, one of the giants of the sea, aboard a Russian whaling ship. So many whales have been caught for their valuable oil that many species are nearly extinct.

Below: This diagram shows how the depth of the sea changes from the continental shelf to the deep ocean.

Bottom: Against a stormy sky, an iceberg drifts slowly north from the frozen continent of Antarctica.

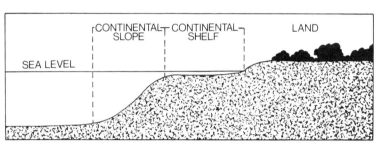

these oceans, however, are a great deal deeper still. These are the great ocean trenches, or deeps, found mostly near land in the Atlantic and Pacific Oceans. The deepest of all is the Mariana Trench, near the island of Guam in the Pacific. It goes down over 11 kilometres.

Water is very heavy, as you will know if you have carried a bucket full of it. The deeper you go in the ocean, the more the column of water directly above you weighs. We say that the water exerts a pressure of so many kilogrammes or tonnes. Aqualung divers cannot dive safely beyond about 30 metres because the water pressure makes it impossible to breathe—though specially trained and equipped aqualung divers have gone further down. In the great deeps the pressure is so great it will crush all but the strongest steel spheres—yet there are some fishes which can live even at these depths, adapted to withstand such pressure.

The temperature of the sea varies very much. On the surface it is below freezing point near the Poles, and in the tropics it can be only a few degrees below blood heat. Deep down the temperature is lower the further you go, until in the great deeps it is only a few

10

degrees above freezing. Because the sea contains much salt, it actually freezes at a far lower temperature than fresh water.

Apart from the great deeps, the bottom of the ocean is far from flat. A large area of it is fairly level, and is known as the abyssal plain. This plain is broken by massive mountain ranges, some of them much larger than any on land. Some of these seamounts are so high that they break the surface, forming islands. Bermuda is the top of a series of mountains.

Many of the mountains under the sea have been made by volcanic action, and you can read more about this on pages 12-13. Many islands, particularly in the Pacific, are the tops of volcanoes—and in some islands of this kind, such as Hawaii, the volcanoes are still active. Some islands, known as atolls, have been formed by coral growing on the top of volcanic peaks. Coral is the limy skeletons of millions of tiny sea animals, which thrive in warm, shallow seas.

The sea is never still. The great waves which rock ships sailing on the sea are up-and-down movements of the water, which has been set moving by several different forces. Earthquakes and volcanoes under the sea start

Coral is formed by the stony skeletons of millions of tiny animals under the sea. These are just a few of the many beautiful shapes you can find.

waves. So do winds, lashing the surface. Another major influence on the oceans is the pull of the Moon and the Sun. This pull, known as gravity, is the same force as the Earth exerts when you drop a brick on the ground (without gravity the brick would stay floating in the air!).

The water of the oceans also moves about in currents. Currents on the surface are set in motion by the winds. There are also currents deeper down which are caused by changes in the temperature of the sea. Surface water heated by the Sun in the tropics flows away, or is evaporated. Cooler water from below rises to take its place.

The oceans are the source of rain. When water is evaporated from the sea in hot parts of the world it is absorbed into the air and is blown along by the winds. The winds themselves are caused by variations in the temperature of the air. As air is heated it rises, and colder air flows in to take its place. When damp air from a warm place reaches colder parts of the world the water in it condenses to form clouds—just as steam from your bath condenses on the cold surface of a mirror. The water in the clouds then falls as rain.

Inside the Earth

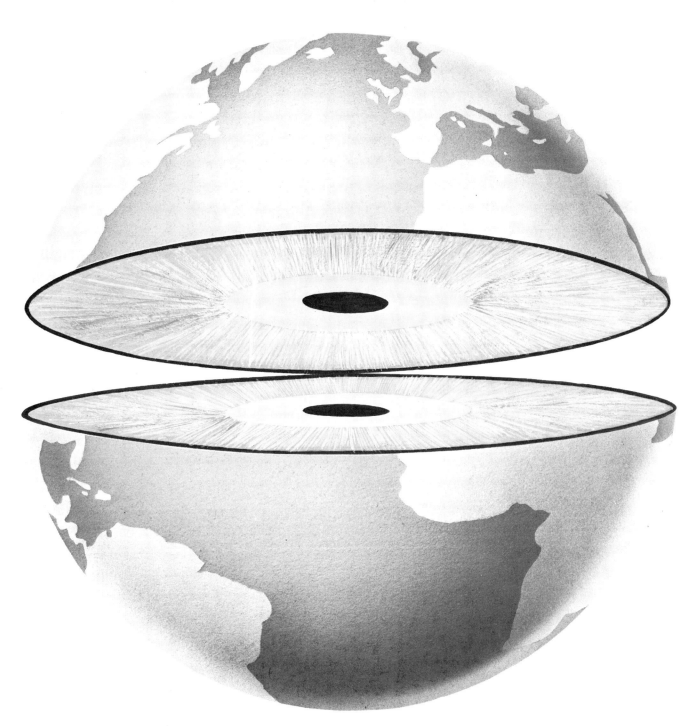

Men have explored nearly all the surface of the Earth. They have ventured into the space around it, and have even landed on the Moon. Yet the one place they have not been able to explore lies right beneath their feet: the inside of the Earth. The deepest that anyone has been into the Earth is the 11-kilometre-deep Mariana Trench in the Pacific Ocean, into which two men descended in a bathyscaphe, a specially-built submarine, in 1960. The deepest hole on land is a mine, only just over 3.5

This is what the inside of the Earth is probably like. The surface of our Earth is very thin and is called the crust. It is shown on this diagram by just a thick black line. Then comes the mantle, made of heavy rocks, followed by the outer core, made of molten iron and nickel, and the inner core, also iron and nickel but in a solid state.

kilometres deep. It is in South Africa, and at that depth the temperature is over 50°C.

However, although the interior of the Earth is so difficult to reach, we know a surprising amount about it. Scientists study the interior by monitoring earthquake shocks. An earthquake in one part of the world is eventually felt in other parts. The shock waves take time to travel, and the time varies according to the kind of material they have to pass through. By taking a great many measurements and comparing them, the scientists have built up a picture of what the Earth is like inside.

The Earth is not a perfect sphere, like a ball. It is slightly flattened at the poles, north and south, so that it resembles a giant orange. Inside it is more like an onion, because it consists of several layers. The outermost layer, the part we all know, is called the crust. The crust is not very thick—as little as 8 kilometres under the oceans, and about four times that under the continents. The crust consists of two layers of rocks. The lower layer is called sima, and its main ingredients are silicon, oxygen, iron and magnesium. Sima extends all over the surface of the Earth, and forms the floor of the mighty oceans.

The continents form the second layer, which lies over the sima. This layer is called sial, and it contains rock made up of silicon, oxygen, aluminium, calcium, sodium and potassium. There is no sial under the deep oceans, but it forms the continental shelf, part of each continent that is under water.

Underneath the crust is the mantle, a thick layer of solid rock. It is nearly half the total distance from the centre of the Earth to the surface. The rock is very hot indeed, and it grows steadily hotter the deeper you go, until it meets the central part of the Earth, the core. The outer core is nearly as thick as the mantle, and it is believed to be iron and nickel, so hot that they are molten.

The inner core is also iron and nickel, but although it is even hotter than the outer core, scientists think it is solid. The reason is that because of the force of gravity, pressures inside the Earth are so great that the atoms of which the inner core is made are packed tightly and the metals cannot flow about.

You can see that the Earth's surface is always changing, as wind and rain and the sea grind away rocks, and river estuaries silt up and become dry land. Changes go on inside the Earth, too. The Earth's crust is always being renewed. It is made up of a series of huge, irregularly-shaped plates, on which the continents sit. In some places these plates overlap. Deep under the Atlantic and Pacific oceans the plates are coming up out of the interior and moving away from each other.

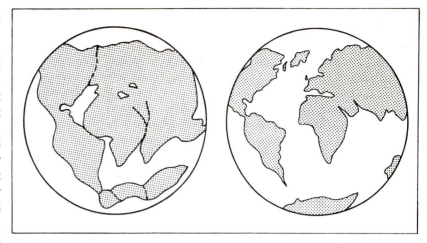

On the left you can see how the continents once fitted together. They moved apart gradually until they took up the positions in which you see them on maps today.

How the sea floor is spreading: new material is added to the plates in mid-ocean, and it is swallowed up under another plate near the coastline. As it goes under the second plate it forces up the land to create a mountain chain.

The Atlantic Ocean is actually growing wider every year, by a few centimetres.

Since the Earth does not get any bigger, the extra crust formed in this way has to go somewhere. So in several places one plate slides under another and goes back into the interior of the Earth. One such place is under the west coast of America. The mountains along that coast have been pushed up as a result of this movement.

You can see how the continents have moved over millions of years if you look at a map. Notice that South America and Africa look as though they ought to fit together. Long ago they did, and scientists think that all the land was one huge mass which broke up and moved around until the continents are where they are now. The presence of coal in Alaska shows that tropical forests once grew there—so Alaska must have been much nearer the Equator than it is now.

Most of the world's volcanoes and earthquakes occur at the places where the plates join. Earthquakes are caused by the edges of two plates rubbing together and moving in sudden jerks. Volcanoes are eruptions of liquid rock from inside the Earth. There is a chain of them along the mid-Atlantic ridge, the line under the ocean where new plates are constantly being formed. Iceland lies on this ridge, and it has many volcanoes. It is also the one island in the world that is always growing bigger. It spreads as America and Europe move gradually further apart.

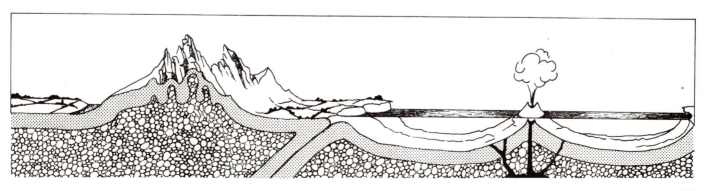

The Moon, Earth's Neighbour

The Moon is Earth's nearest neighbour. It travels round the Earth once every 27 days 8 hours, in a path that is not a true circle. At its nearest it is only 363,297 kilometres away. At its furthest it is about 42,000 kilometres further away. The nearest any of the planets gets to Earth is about 39,500,000 kilometres, reached by Venus when it comes between us and the Sun.

At one time people used to think that the Moon was a part of the Earth that had been torn away. Now scientists believe it is a small planet which has been 'captured' by the Earth. The capture came when it strayed into the Earth's gravitational field, the force which pulls things towards the Earth. It is this force which keeps the Moon in orbit around the Earth, just as a piece of string will keep a weight on the end of it from flying away when you whirl it around your head.

The Moon also has a gravitational pull, as indeed all heavenly bodies do. The Moon's gravity is only one-sixth as strong as that of the Earth, because it is much smaller. However, the Moon's gravity is strong enough to affect Earth, and is the cause of tides in the sea. It pulls the sea towards it on the side of the Earth which it is facing, and also pulls the solid mass of the Earth slightly away from the

Opposite: The Moon as we see it from the Earth, photographed through a giant telescope.

The dry, craggy surface of the Moon, photographed from a spacecraft about 110 kilometres above it. You can see some of the many small craters which pockmark the Moon, the result of a ceaseless bombardment of meteorites from outer space.

sea on the opposite side. The Sun also affects the tides, but because it is so much further away its gravitational force is less than half that of the Moon.

The Moon produces no light of its own. Instead it reflects the light of the Sun like a huge mirror. The position of the Moon in relation to the Earth determines how much of the lit surface we see. When the Sun is shining from behind the earth we can see the whole disc lit up—the full Moon. When the Moon is between the Sun and the Earth only the dark side faces us, and the Moon is invisible. As the Moon moves round, a thin crescent of light appears—the new Moon. Gradually the visible part grows until full Moon is reached. Then it shrinks again until the Moon is invisible once more.

We only see one side of the Moon. This is because the Moon spins on its axis very slowly, revolving once in the time it takes to go right round the Earth.

The Moon is the only heavenly body that Man has so far been able to visit (see pages 24-25). From these visits and from information sent back by unmanned spacecraft we now know a great deal about it. First of all, the Moon has no atmosphere worth speaking of. The Moon's gravity is too weak to keep the gases that make up an atmosphere. Earth visitors have to take their own air supply and breathing apparatus with them. Another effect of having no atmosphere is that the sky is always black, even in broad daylight. Earth's blue sky is produced by the molecules of air scattering the rays of light from the Sun.

Earth's atmosphere acts as a great, invisible shield. It not only filters out harmful rays in sunlight, but also wards off bombardment by meteors, small lumps of rock that are found in space. The friction of the air on the meteors makes them so hot they burn up before they reach ground. Only a few meteors get through this air shield. The Moon has no such protection. Its dry, dusty face is pockmarked with craters from meteors which have struck it over millions of years.

The surface of the Moon has changed very little for a long time. Because there is no atmosphere, there is neither rain nor wind to wear away rocks and blow things about.

Without air and water there is no life on the Moon. Its surface is dead. Yet deep inside the Moon is not dead. Moonquakes show that there is volcanic activity and therefore the interior is probably very hot—though not so hot as the Earth. At night the Moon is lit by earthshine, the Sun's light reflected from the Earth just like moonshine.

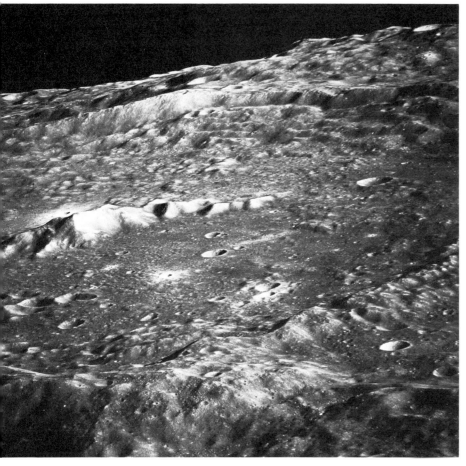

The Wandering Planets

Our Earth is just one of a group of heavenly bodies which are in orbit around the Sun. They are called the planets, a word which comes from a Greek word meaning a wanderer. Ancient astronomers gave them this name because they appeared to wander about the sky, while the stars stayed in the same places in the heavens.

There are nine planets. Mercury is nearest to the Sun, and then follow in order Venus, Earth, Mars, Jupiter, Saturn, Uranus, Neptune and Pluto. Between Mars and Jupiter are the asteroids, a group of very small planets.

There are more than 1,600 of them, and the biggest, Ceres, is only 687 kilometres in diameter. Some of the smallest are only 1 kilometre or so across. They may be the remains of a large planet which broke up.

The planets vary enormously in size, as you can see in the table on this page. To give you a more vivid idea, if Earth were the size of a marble Jupiter would be about as big as a football. Mercury, Mars and Pluto are smaller than Earth; Venus is similar in size, while Saturn is nearly as big as Jupiter. Uranus and Neptune on the same scale would be about the

Right: Venus photographed from the U.S. space probe Mariner 10 at a distance of 2,780,000 kilometres. Far right: Mercury, also photographed from Mariner 10. The distance this time was 3,500,000 kilometres.

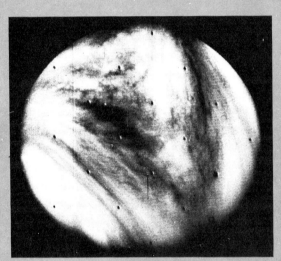

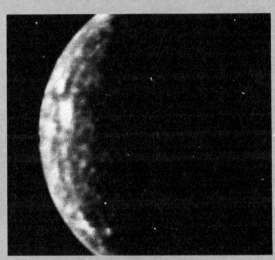

VENUS: Maximum distance away from Earth 260 million kilometres. Minimum distance from Earth 40 million kilometres. The spacecraft Mariner 10 which sent us back this photograph took 98 days to reach Venus.

MERCURY: Maximum distance away from Earth 217 million kilometres. Minimum distance from Earth 82 million kilometres. Mariner 10 which sent back this photograph took 142 days to reach Mercury.

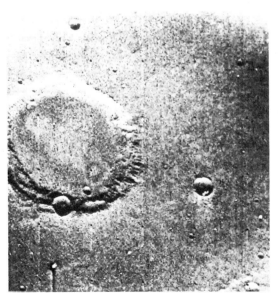

One of the many craters on Mars, photographed by the space probe Mariner 6.

size of a croquet ball. Like Earth, each planet orbits the Sun and also spins on its own axis. None of the planets produces its own light, but they reflect the Sun's light, just as the Moon does.

Mercury is the smallest of the planets, and the nearest to the Sun. It turns very slowly, so its day lasts nearly 59 of our days, but its year, the time it takes to go round the Sun, lasts only about 88 Earth days. It has no atmosphere. Photographs taken by space probes show that it has a dry, dusty surface, very like that of the Moon.

Venus, next from the Sun, has a day lasting 243 of our days, and its year is shorter than its day—225 Earth days. It has an atmosphere, a poisonous one consisting mainly of carbon dioxide. The temperature on its surface is hot enough to melt lead, and life could not exist there.

Mars has a day only a few minutes longer than that of the Earth, but its year is nearly twice as long, at 687 Earth days. It has a very thin atmosphere, mostly carbon dioxide, but also with some water vapour in it. Its surface has great valleys which appear to have been carved out by the action of water, just as valleys are made by rivers on Earth. Most of the surface is dry and dusty, with no trace of water, but there are ice caps at the poles. The ground is permanently frozen because the temperature never rises above freezing point, so there may be water trapped in the soil.

Jupiter is the largest of the planets. Its day

THE PLANETS	
Name	Diameter in kilometres
Mercury	4,990
Venus	12,400
Earth	12,757
Mars	6,790
The Asteroids	1 to 687
Jupiter	143,000
Saturn	121,000
Uranus	52,000
Neptune	48,000
Pluto	5,900

Right: This general picture of Mars was taken by a U.S. Viking space probe. Note the band of haze, showing that the planet has an atmosphere. Far right: Jupiter photographed from a U.S. Pioneer 10 space probe. The Great Red Spot shows up very clearly.

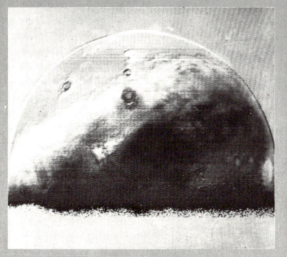

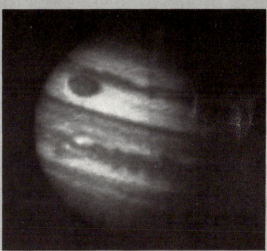

MARS: Maximum distance away from Earth 101 million kilometres. Minimum distance away from Earth 56 million kilometres. The spacecraft Viking 1 which sent us back this photograph took 334 days on its voyage to reach Mars.

JUPITER: Maximum distance away from Earth 966 million kilometres. Minimum distance away from Earth 591 million kilometres. The spacecraft Pioneer 10 which sent us back this photograph took 637 days to reach Jupiter.

lasts only about 12 of our hours, but its year is equal to almost 12 Earth years. Although it is more than 11 times as large as the Earth, it is made of much less solid materials. It has an atmosphere, consisting largely of hydrogen, the lightest of all gases. So far astronomers have not been able to see through this atmosphere to find out what the surface is like, or whether the whole planet is made of gases. The planet has a number of coloured bands around it, and several dark and light spots. The largest is called the Great Red Spot.

Saturn, the second largest planet, has a day lasting about 10 of our hours, and a year lasting more than 29 Earth years. Its deep atmosphere is mostly helium, hydrogen and methane. The surface is probably solid because its temperature is −155° C. Saturn is

surrounded by a series of rings only 16 kilometres thick. These rings are probably made up of ice crystals.

Uranus has a day lasting just over 10½ of our hours, and a year 84 Earth years long. Unlike the other planets, as it spins it lies over on its side. It appears to have an atmosphere, but we know nothing about its surface.

Neptune, which like Uranus is one of the larger planets, has a day lasting about 14 Earth hours, and a year about 165 Earth years long. Astronomers discovered it in 1846. It seems to have an atmosphere containing several gases, including methane, ammonia and hydrogen.

Pluto, the furthermost planet, was found in 1930. Its day lasts just over six of our days, and its year is 249 Earth years long. It appears to have an atmosphere.

The Life-giving Sun

Life on Earth could not exist without the Sun. It provides warmth: without it the Earth would be freezing cold. It provides energy: plants use the energy in sunlight to help them make their food, and we use that energy, stored by plants many millions of years ago in the form of the fossil fuels, oil and coal. The Sun also provides light, which is a form of energy.

Earth and the other planets are in orbit around the Sun. So far as we know Earth is the only planet on which conditions are right for life to exist, but there may be others, because the Sun is a star, one among many millions of stars in the Universe. It is a typical dwarf star, that is, one of the smaller stars. Do not be misled by the word 'small': the Sun has a diameter more than 100 times as big as that of the Earth, and more than a million Earths

This photograph of the Sun was taken on July 13, 1937 at a time of maximum sunspot activity. You can see more than 20 groups of sunspots in the picture.

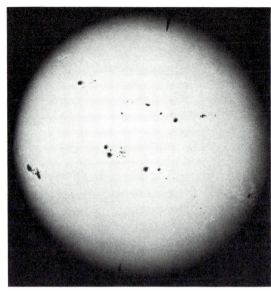

Below: A huge flare, an irruption of radiation from the Sun. Such flares rarely last more than 20 minutes. This one was on July 16, 1959.

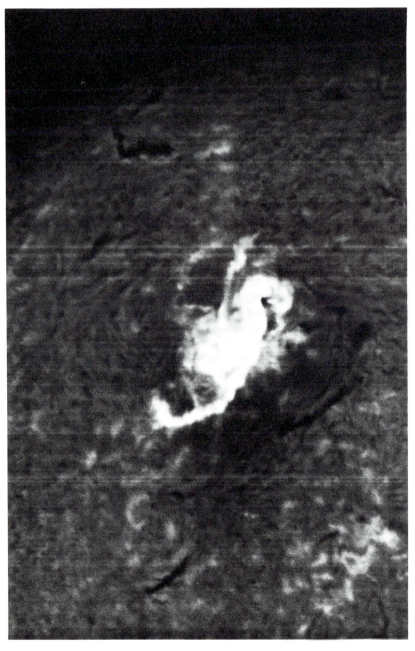

would fit inside it. However, many stars are thousands of times larger than the Sun.

The Sun is about 150,000,000 kilometres away from us. Its light takes about eight minutes to reach Earth. By studying that light scientists have been able to find out a great deal about the Sun – what chemical elements it is made of, for instance. You should never look directly at the Sun, even through strong smoked glasses. Although it has travelled such a long way, its light is powerful enough to send you blind. However, by taking precautions astronomers are able to watch what is going on in that fiery globe. They can learn more during an eclipse (see page 15), because when the Sun's disc is hidden by the Moon the cloud of burning gases around it can be seen more clearly. Even better observations are obtained from orbiting space laboratories.

Although most of the known elements have been detected in the Sun, the most common are hydrogen, helium, calcium, sodium and iron. Hydrogen provides the fuel for the Sun's heat. Our star is a giant nuclear furnace, in which hydrogen atoms, 4,000,000 tonnes of them every second, combine to form the slightly heavier gas helium and in doing so release vast quantities of energy.

This energy pours out of the Sun in a steady stream. A tiny part of it reaches Earth and the other planets. The rest escapes into outer space. Yet scientists calculate that the Sun's resources are so great that it can continue to produce this tremendous outflow of energy for thousands of millions of years to come. From time to time a kind of energy storm happens on the Sun, and the output of radiation is suddenly and briefly increased. A storm of this kind struck Earth in November 1960. There was nothing to see or hear, but the

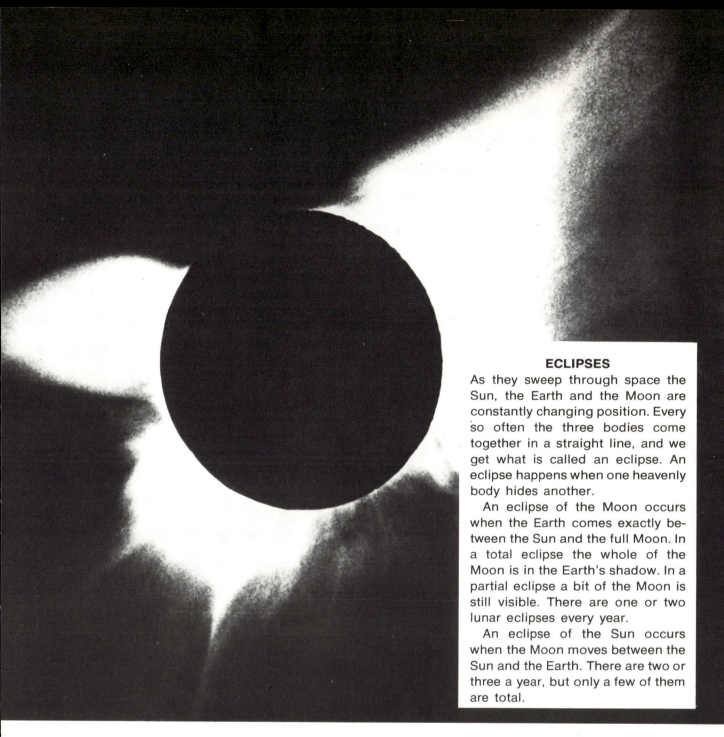

ECLIPSES

As they sweep through space the Sun, the Earth and the Moon are constantly changing position. Every so often the three bodies come together in a straight line, and we get what is called an eclipse. An eclipse happens when one heavenly body hides another.

An eclipse of the Moon occurs when the Earth comes exactly between the Sun and the full Moon. In a total eclipse the whole of the Moon is in the Earth's shadow. In a partial eclipse a bit of the Moon is still visible. There are one or two lunar eclipses every year.

An eclipse of the Sun occurs when the Moon moves between the Sun and the Earth. There are two or three a year, but only a few of them are total.

electrical disturbance blacked-out long-distance radio, upset the needles of compasses, and caused electric lights to flicker.

The Sun itself sends radio signals, and these signals vary in strength. The variations, and also the magnetic storms just described, appear to be linked with dark patches which appear from time to time on the surface of the Sun. They are known as sunspots. Some last several weeks, others only a few days.

The energy which radiates from the Sun reaches Earth in a series of rays of different wavelengths. Many of these rays are harmful to life, but fortunately the Earth's atmosphere blocks most of the dangerous ones. The atmosphere lets light rays through, and some of the infra-red rays, which give us heat. It also allows in part of the ultra-violet rays, which are the ones which cause sunburn. The rest of the ultra-violet rays, plus X-rays and

An eclipse of the Sun. The whole of the Sun's disc is covered by the Moon, and you can see the vast sheets of burning gases shooting from the surface.

cosmic rays, are bounced back into space by the atmosphere. However, men orbiting the Earth and travelling to the Moon in spacecraft could be affected by these rays.

For many years people believed that the Sun moved while the Earth stood still. We now know that the daily passage of the Sun across the sky is caused by the movement of the Earth, so we are looking at the Sun from a series of different viewpoints. In the same way the position of a street-light appears to change as you drive past it in a motor-car.

However, the Sun is also moving. For one thing, it is turning on its axis, just like the Earth and the planets. It takes about 25 days to revolve once.

The Sun – and all the planets with it – is also rushing through space at the rate of 19 kilometres a second. Other stars in the heavens are moving with it.

The Distant Stars

If you look at the sky on a clear night you can see hundreds of stars, tiny points of light twinkling in space. Look through a small telescope and you will see many more. Look through a large telescope, and even greater numbers spring into view.

Each star is a sun, producing its own heat and light by nuclear reaction. Although they look so small, many stars are much bigger than the Sun; their tiny appearance is because they are so far away. The nearest star beyond the Sun is called Alpha Centauri. Light from that star takes $4\frac{1}{3}$ years to reach us, while light from the Sun takes only eight minutes. Even this huge distance is small in terms of space—other stars are thousands and millions of times further off.

There are many different kinds of stars, and we can tell the differences by the kind of light and other radiation they produce. Supergiants are the biggest stars, and they have diameters several hundred times that of the Sun. However, the material of which they are made is much less dense, and they consist mostly of gases. Below these come giants and subgiants, and then the normal-sized stars. Next in size come dwarfs, of which the Sun is one; they are much smaller than normal-sized stars. White dwarfs are a special kind of star which are about the same size as the Earth. These white dwarfs are very dense indeed, and the matter of which they are made is much heavier than anything found on Earth.

Other variations in stars are seen in their brightness. Some are very hot—about 100,000 times as hot as the Sun. Others are comparatively cool, and give off mostly infra-red rays. Some stars appear to change in brightness, becoming brighter and then duller every few days. They are called pulsating stars. Double stars are pairs of stars that circle around each other, like two boxers sparring for an opening.

Sometimes new stars appear. They are called *novae*, from the Latin word for new. They are not really new stars, but stars which have suddenly become brighter because a huge explosion has taken place. *Novae* stay bright for a while, and then gradually fade.

For hundreds of years astronomers have grouped stars together according to patterns they make in the sky. They call these patterns constellations. Familiar constellations are the Plough (also called the Big Dipper), which you can see from the Northern Hemisphere, and the Southern Cross, which people see in lands south of the Equator.

Constellations are merely patterns as we see them, but stars do form vast groups called galaxies. Our Sun forms part of the Milky

Way (sometimes called The Galaxy), which contains thousands of millions of stars. The Milky Way is like a giant wheel of stars in the sky, rotating around a central point, the nucleus. The Sun is about one-third of the distance from the centre to the edge of the Milky Way, and it orbits the centre once in about 250,000,000 years.

For long enough people thought there was only one galaxy, but the American astronomer Edwin Hubble discovered in the 1920s that the fuzzy clouds which other astronomers had detected far out in space were not clouds of gas but other galaxies. There are several thousand million of these galaxies so far identified.

Hubble also found that the Universe is steadily stretching, for the galaxies are moving apart from each other at enormous speed. The further off they are, the faster they are going, and the speed is always increasing. The most distant objects we have detected so far are known as quasers, which is short for 'quasi-stellar radio sources'. They were first detected in the 1960s as mysterious sources of radio waves, far out in space, which appeared to be something like stars. When astronomers studied the spectra (rainbow-like breakdowns of light) from the quasers they found familiar patterns of lines on them that indicated the substances of which they are made, generally hydrogen. However, the lines appeared not in their usual place in the spectrum but shifted towards the red end of it. This red shift, as it is called, means that the object is moving away from us. It is called the Doppler effect, and you can hear the same effect in sound with the siren of a police car. As the car rushes away from you, you can hear the pitch of its notes getting steadily lower.

Earth and the Universe Inquiry Desk

What makes a cave? (see below and right)

Nearly all caves are made by running water. Caves along the coast are formed as the sea ceaselessly pounds away at the rock. If part of the rock is softer than the rest the sea cuts through it and forms the cave (A). The rocky arches you can see off some coasts are the remains of caves where the sea has washed away all the rest of the rock, leaving only the very hardest part.

Land caves are nearly always found in limestone or similar soft rocks. Water seeps through cracks in the rock, dissolving some of the limestone (C, D, E). Over millions of years the water hollows out passages and rooms. Some caves are now completely dry, while others, such as Labouiche in southern France, have rivers running through them. A few caves are formed in the lava from volcanoes: as the outer layer of lava cools and becomes solid the molten lava underneath drains away leaving a cavern (B).

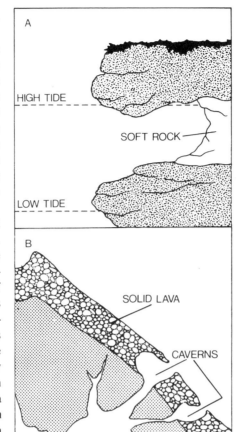

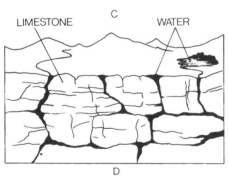

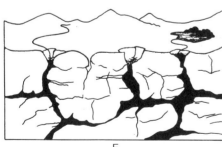

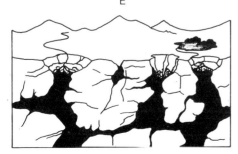

? ? ?

Why is an artesian well so-called? (see below)

Artesian wells are named after the province of Artois, in France, where these wells were first made. An artesian well is made in areas where a layer of porous rock, which will hold water, lies between two layers of fairly watertight material, such as clay. The bed of water-holding rock lies along the line of a slope, and at the top of the slope it comes to the surface. Rain falling on the exposed rock soaks through it. A well bored lower down the slope produces a gush of water.

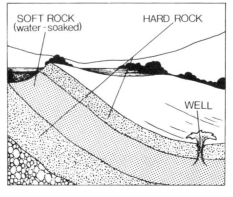

? ? ?

How did fossils get into the ground? (see below)

Fossils are the remains of animals and plants which lived long ago. If dead things are quickly buried, by soil or mud piling over them, they are protected from rapid decay. In time the soft parts do decay, but the harder parts may survive. Sometimes minerals in the soil penetrate into spaces in bones and shells, and so help to preserve them, partly as stone and partly as the original material (1). More often the original bone and shell dissolve away and are replaced by minerals, which make a perfect replica of the missing parts (2). Some plants and animal bodies are trapped in a substance such as clay which hardens around them. The bodies decay leaving hollow moulds of the originals (3).

? ? ?

What causes waves on the sea?

Waves are caused by the wind blowing on the surface of the sea. In just the same way, if you blow through a straw on to the surface of water in a glass you make ripples on it. The waves move forward, but the water does not: when a piece of wood or a seagull floats on the surface it bobs up and down over the same patch of water, without moving forward.

The reason is that the wave movement is like the wave in a piece of rope when you tie one end to a fixed hook and move the other end up and down. The wave moves along the rope, but the particles of the rope stay in the same place. The sea does move to and fro, but this movement is due to tides (see pages 14-15).

? ? ?

What is the difference between climate and weather?

We use the term 'weather' to describe what is happening in the atmosphere at a particular place and time: if it is dry or wet, cold or hot, stormy or peaceful. Climate is the average weather found in a particular place over the whole year. So a place with a warm, dry climate may have days when the weather is actually cold and wet. Near the Equator the terms climate and weather tend to mean the same thing because the weather does not change very much.

? ? ?

What is a light year?

It is the distance that light travels in one year, and it is used for measuring the vast distances in space. As light travels at nearly 300,000 kilometres a second you can see that a light-year is a very long distance indeed! Only about 20 stars are within 12 light-years of the Earth.

? ? ?

Do black holes really exist? And if they do what are they?

We do not know for sure if there are such things as black holes, but many astronomers think there must be because there are sources of energy in outer space that they cannot account for otherwise.

The theory of a black hole is based on the behaviour of gravity. An object that is pulled by gravity—such as a falling ball—picks up energy from its motion through space. In a black hole the pull of gravity has become so strong, and the energy produced by objects falling into it is so great, that even light cannot withstand its pull. So no information—light, radio signals or anything else—can leave a black hole. Some astronomers think that stars as they collapse in on themselves in cooling eventually end up as black holes. At some point the process goes into reverse, and we get a white hole, pouring out energy.

? ? ?

What are the Northern Lights?

The Northern Lights, or Aurora Borealis as they are also called, are bands of brilliant colours in the sky which can be seen in the polar regions of the Northern Hemisphere. They are caused when streams of particles from the Sun, deflected by the Earth's magnetic field, penetrate the atmosphere. There they collide with molecules of air and make them glow. Similar lights seen near the South Pole are called the Aurora Australis. See the picture below.

? ? ?

Do comets foretell disaster?

For many thousands of years people used to think that comets were evil omens, and that following their appearance terrible disasters would take place on Earth. Every so often one appears in the sky, and with its streaming tail it looks unlike any other heavenly body. So people were naturally worried by comets.

Today we know that a comet is just a heavenly body travelling in a long, curved path around the Sun. The head or nucleus of a comet is probably made up of a mass of frozen gas and dust, and it glows in the sunlight, like the Moon and the planets. When the comet gets near the Sun the heat turns the outer layers of gas and dust into vapour so it becomes much larger and fuzzier. At the same time some of the gases spread out from the head to form a glowing tail. The tail is caused by radiation from the Sun, and always points away from the Sun. A disappearing comet seems to be moving backwards.

The most famous comet—Halley's Comet—was named after the astronomer Edmund Halley, who plotted its path in 1682 and predicted that it would return in 1758—which it did. The comet last appeared in 1910 and is due again in 1986.

Exploring Space

Men have been fascinated by the heavens for thousands of years: even cavemen drew charts of the stars, showing keen observation. Ancient astronomers had two things which today's observers find hard to get: skies free of industrial pollution and plenty of time. Early on they discovered that while the stars appeared to revolve around the Earth in a fixed pattern, some heavenly bodies—the planets—moved about. Some not only went round the sky, but even seemed to go backwards from time to time.

As a result of hundreds of years of observations, carefully noted, astronomers built up an amazingly detailed and accurate picture of space beyond the Earth. For long enough most people thought that the Earth was the centre of the universe, with the Sun, Moon, planets and stars revolving around it, although about 270 BC the Greek astronomer Aristar-

The first man in space—Russian cosmonaut Yuri Gagarin just before his epoch-making flight. He was then 28 years old. Gagarin was killed in a plane crash in 1968.

Below: Stepping on to the Moon: Edwin Aldrin becomes the second man to set foot on the Moon, photographed by the first man, fellow astronaut Neil Armstrong.

chus thought it possible that the Earth revolved around the Sun. His ideas were revived in 1543 by the Polish astronomer Nikolaus Copernicus.

Before the invention of the telescope men had to depend on what they could see with the naked eye. For example, they knew only of five other planets: Mercury, Mars, Venus, Jupiter and Saturn. The first astronomer to use a telescope was the Italian Galileo Galilei. Not only did he see many more stars, but he discovered the satellites of Jupiter and found that the Moon had a rough surface.

For more than 300 years after Galileo died in 1642 men continued to explore space by a mixture of observation and mathematical calculations. They discovered the planet Uranus in 1781, and from the way it moved guessed that there must be another planet whose gravitational force was affecting it. Gravity is the pull of a body: the Earth's gravity pulls things towards it, so that if you let go of something such as a ball it falls to the ground. In the same way the pull of the Earth keeps the Moon in orbit round it.

This second planet, Neptune, was found in 1846, and it too appeared to be 'pulled' by some unknown body. The last planet to be found, Pluto, was discovered in 1930.

So far all this space exploration had been carried out from Earth, but many men dreamed of the day when they could actually leave Earth and travel into space. This dream became a reality when scientists discovered how to make and operate huge rockets. You can read about the way in which rockets work and Man can travel in space on pages 186-187.

The first big step forward in space exploration came on October 4, 1957, when the Russians launched the first artificial satellite, *Sputnik 1.* It weighed just 83.6 kilogrammes and orbited the Earth once every 96 minutes. A month later the first space traveller left Earth, a small dog named Laika, aboard *Sputnik 2.* The first American satellite, Explorer 1, was launched on January 31, 1958.

The first man in space was the Russian Yuri Gagarin, who made an orbit round the Earth on April 12, 1961. During the next few years the Russians and Americans made a great number of space flights, some manned, others not. The first spacecraft to land on the Moon was sent there by the Russians in 1966. It radioed pictures of the Moon's surface back to Earth. Earlier space probes had hit the Moon, destroying themselves, or had gone into orbit around it, taking photographs.

The Americans began in 1961 a costly programme to put a man on the Moon by 1970, and they succeeded on July 20, 1969, when two U.S. astronauts, Neil Armstrong and

Orbiting the Moon: the command and service modules of Apollo 17, the last manned mission to the Moon. This picture was taken from the Lunar Module 'Challenger' as it returned from the Moon to dock in space with the main part of the spacecraft.

Edwin 'Buzz' Aldrin, set foot there. For this and other Moon voyages the Americans used a series of Apollo spacecraft. The craft each carried a landing module which could be detached from the main spacecraft and land on the Moon's surface, while the rest of the Apollo flew in orbit around the Moon. Altogether the Americans made six Moon landings, the last in December 1972.

Space exploration today is concentrating on sending unmanned spaceprobes to fly past or land on the planets. From these we have learned that Mars has a surface much like that of the Moon, and that Venus is unpleasantly hot. Both the Russians and the Americans are concentrating on endurance flights in orbit around the Earth, in the hope that a space station can be constructed in orbit to allow easier exploration of outer space.

Meanwhile exploration of space from the Earth's surface continues. Modern radio telescopes are picking up information about stars that are too distant for ordinary telescopes.

The Living World

You all know that you are alive, but what is it that makes you different from the rocks below, or from a clockwork doll? If you look at the range of living creatures in the world today, you find that they have certain things in common. Perhaps the most obvious point is that all living creatures reproduce themselves. Humans have babies, cats have kittens, worms produce eggs that hatch into tiny worms, and plants make seeds.

Living things also grow: they increase in size by producing more of the living matter of which they are made. In order to grow, they must take in the substances they need to make living matter.

All living creatures respond to their surroundings. Animals are usually more active in warm weather—unless it is very hot—than in cold weather, and are attracted to move toward certain smells and away from others. Plants bend towards the light. The length of the days and the temperature cause plants to flower at certain times of the year, and make birds migrate from one country to another.

Living things are classified as members of the plant or animal kingdoms according to how they feed. Plants make their own food. Water and dissolved minerals from the soil enter the plant through its roots and travel up the stem in tiny tubes to the leaves. There they are combined with carbon dioxide gas from the air, which enters the leaves through tiny holes to make the more complex substances that go to form the body of the plant.

To build these materials the plant needs energy. It gets this from sunlight, which is absorbed by a green pigment called chlorophyll in the leaves. So the plant makes all the substances it needs to build its own tissues. The process is known as photosynthesis, from Greek words meaning 'putting together with the help of light'.

The substances also provide the plant with an energy store. Just as energy has to be put in to make the complex body substances so, if these substances are broken down into simpler materials, the energy trapped in them is released.

The process by which substances are broken down to release energy is called respiration. Both plants and animals respire. Oxygen from the air is used to burn up the substances, and carbon dioxide gas is given off in the process. This is why you need to breathe—oxygen is taken in through the lungs as you breathe in, and carbon dioxide is given off when you breathe out.

Animals cannot make their food: they have to take in ready-made food in the form of

A typical plant cell, with a strong cell wall made of cellulose. The chloroplasts contain chlorophyll, which makes a plant green and helps it to use the energy of sunlight to make its own food.

Below: How water and minerals travel up the stem of a plant to its leaves in tiny tubes (vascular bundles).

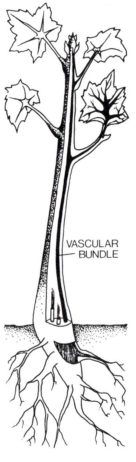

VASCULAR BUNDLE

A typical animal cell, which is surrounded by a fine membrane instead of a tough cell wall. In the cell is a watery substance called cytoplasm, and in the cytoplasm are a number of tiny bodies called organelles.

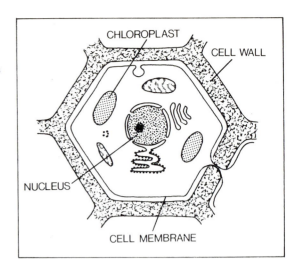

CHLOROPLAST
CELL WALL
NUCLEUS
CELL MEMBRANE

plants or other animals. This they use to build their body structures and to obtain energy by respiration.

The living matter which makes up plants and animals occurs in small units called cells —tiny bags of living matter surrounded by thin skins called membranes. From these tiny building blocks all the body tissues are built. Each cell has a nucleus, a little bag containing the information that tells the cell what to do and when. The nucleus is the control-house.

Plant and animal cells differ in several ways. Plant cells contain the devices for making their food. These are the chloroplasts, which hold the green pigment for trapping sunlight. In the middle of each plant cell is a bag filled with water, and round the outside of the cell is a tough elastic cell wall. The bag fills up with water until the cell wall can stretch no more. Then the cell is stiff, rather like a bicycle tyre after you have pumped air into the inner tube. Large plants such as trees have woody material for extra stiffening.

Animal cells do not have chloroplasts, and they have no tough cell wall. Instead, animals have a skeleton for support, either a bony internal skeleton, as humans have, or a hard skeleton on the outside, as an insect has.

There are some living forms which do not

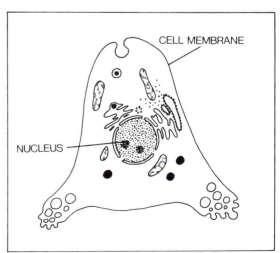

CELL MEMBRANE
NUCLEUS

quite fit into the plant or animal kingdoms. They are bacteria. Bacteria are minute one-celled creatures, sometimes with a cell wall, sometimes not. They have no nucleus. The controlling information is loose in the cell. Some bacteria use sunlight to make their own food, as plants do, while others live on the tissues of dead plants and animals.

Viruses are even more difficult to classify and there has been much discussion as to whether they are living or not. A virus is simply a piece of genetic (hereditary) information wrapped up in a protein coat. It cannot feed, breathe, grow or reproduce unless it finds its way into a living cell. When it does so its genetic information takes control of the cell and uses it to make all the virus's requirements and create new viruses.

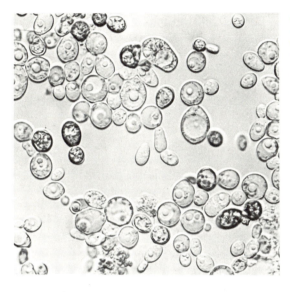

Left: Brewer's yeast, which is used for making beer. It is a kind of fungus, and one of the simplest types of plants. This picture has been magnified 700 times.

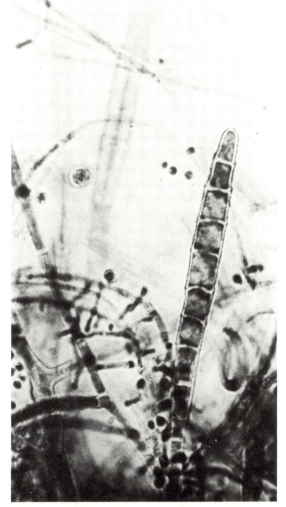

Left: Another kind of fungus. This one causes the common complaint known as athlete's foot. The picture has been enlarged 1,200 times.

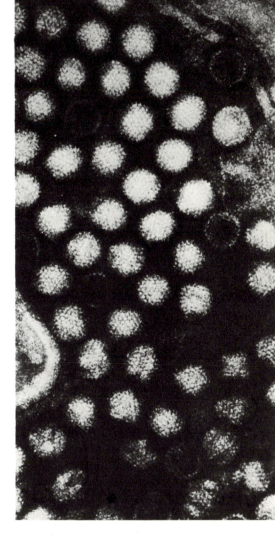

Right: This is a virus, the one which causes infective hepatitis—an inflammation of the liver which produces jaundice. The picture was taken with an electron microscope and is enlarged 150,000 times.

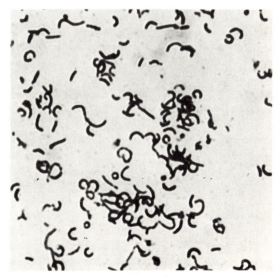

Right: The bacteria *Vibrio comma*, which cause the disease cholera. This picture has been enlarged 3,000 times.

The Plant Kingdom

Would you recognise a swimming plant if you saw one? Some of the tiniest members of the plant kingdom are just single floating cells with two whip-like hairs that beat the water to propel the plant along. Like most other plants they are green, they make their own food, and their cells are surrounded by elastic cell walls. These primitive plants are the algae, and they include all the green plants without obvious stems and leaves. Not all algae are as simple as these swimming forms. Larger algae include the many-coloured seaweeds.

The simplest plants to show a stem and root structure are the liverworts, relatives of the mosses. Liverworts grow in very wet places. Some liverworts look rather like leafy plants, but if you examine them closely you will find that they do not have separate leaves, but merely deep folds in the frilly plate that makes up the liverwort.

The mosses are the simplest leafy plants. They have tiny branched hairs for roots, a slender stem and many leaves. They, too, usually live in shady places, and like the

Above: Cushions of white fork moss growing underneath pine trees.

liverworts they reproduce by spores. Spores are tiny round structures with a hard coat, inside which are contained the first few cells of a baby plant. When the spores are ripe the spore-case opens and the spores blow away on the wind to find new homes.

The ferns are more advanced plants than the mosses. Running up the stems and into the leaves are a series of tiny tubes, some of which carry water and dissolved minerals from the roots to the leaves. Others carry food from the leaves to the roots in summer, and stored food from underground stems and roots to the young growing shoots in spring. Ferns also reproduce by spores.

The conifers or gymnosperms—plants or trees with needle-like leaves and cones instead of flowers—are much more complicated plants. The cones are whorls of woody leaves, on the back of which are little bags containing the reproductive cells. Some contain female cells, attached to the spore-leaf by tiny stalks. Others contain male cells—pollen grains.

When they are ripe the pollen grains blow away on the wind as a cloud of yellow dust. If one pollen grain lands on a female spore-leaf it combines with the female cell, and a little conifer starts to develop. The spore-case hardens around this tiny plant and develops a long wing, and eventually the seed blows away on the wind.

The flowering plants or angiosperms are the most successful plants on Earth, being found in most places, including ponds and streams.

Edible mushrooms are the most familiar form of fungi. The part you see is the fruiting body of the plant. The rest of it is growing under the soil.

The flowers are the reproductive parts of the plants. If you study a flower closely, starting at the outside, you will first find a ring of tiny green leaves—the sepals—which protected the flower while it was still a bud. Then come the larger, brightly-coloured petals, designed to attract insects to the flower, and often having nectaries at their base containing a sweet-scented sugary liquid—nectar. Next come rings of stamens, tiny stalked bags containing the yellow pollen grains. In the very centre is the round seed-box, and on top of it a stalk with a sticky knob, the stigma.

In most cases insects carry the pollen from one flower to another. While the insects are probing for nectar, pollen from the stamens brushes on to their bodies. When the insects reach the next flower the pollen rubs off them, and sticks to the stigma. It makes its way down to the seed-box, where it combines with the female cells. The seed which develops contains a big store of food, and it is shed as a fruit. A few kinds of flowers, such as the grasses, are wind-pollinated like the conifers.

Also included in the plant kingdom are the fungi. Mushrooms and toadstools and the green mould on bread are all fruiting bodies of fungi, designed to release millions of spores into the air.

Fungi do not make their own food like most plants. They produce juices which dissolve the rotting organic material on which they live, then absorb this liquid food.

The wind dispersing the seeds of a dandelion. Each seed is attached to a spray of 'down' which acts as a parachute.

Below: Close-up of an open tulip flower, showing the stigma in the centre on top of its stalk, the style, with six stamens around it carrying the pollen.

The Animal Kingdom

The animal kingdom includes a remarkably varied array of creatures, ranging in size from the microscopic, one-celled, shapeless amoeba to the gigantic blue whale, over 30 metres long and weighing as much as 1,500 men.

The simplest group of animals is the Protozoa (proto=first, zoa=animals). Each animal is a single cell, some with a definite shape, some like a shapeless bag of jelly. Billions of protozoans float in the surface waters of oceans and lakes. Others are parasites in the bodies of larger animals.

A sponge is really a community of different kinds of single-celled animals, all living together in a branching vase-like skeleton of lime or glassy material which they make themselves.

Another group of underwater animals is the coelenterates, animals whose bodies are like hollow cups, with a mouth at one end surrounded by a ring of stinging tentacles. Some, such as hydras and sea-anemones, live attached to rocks or plants. Others, such as jellyfish, swim about.

Annelids are a group containing the earth-

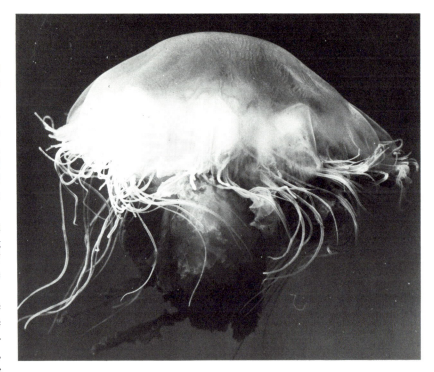

worms and the seashore lugworms, whose coiled casts are a common sight on muddy shores at low tide. Annelid worms have long, cylindrical segmented bodies.

Molluscs are a group containing the shellfish —mussels, cockles, limpets and oysters—as well as snails, octopuses and squid. All molluscs have a foot-like part of the body and a hump containing the digestive parts, which is usually protected by a shell. They breathe in water by means of gills, flat sheets of tissue through which oxygen is absorbed, except for the land snails and slugs which have lungs.

The arthropods (=jointed limbs) make up a big group. They include crustaceans, mil-

lipedes, centipedes, spiders and their allies and insects. All have jointed legs and a hard outer skeleton. As an arthropod grows it sheds its outer skeleton at intervals, the new one stretching to accommodate the larger animal before it hardens.

The crustaceans, which include crabs, lobsters, shrimps, water-fleas and woodlice, have feelers, biting jaws and a shield-like plate protecting their backs. Millipedes and centi-

Above: A lion's mane jellyfish, whose sting can cause a severe weal on the skin. Far left: The common whelk, a kind of mollusc.

The pelican is a large bird that lives in and near water. It has a large bill with an elastic pouch, which it uses for scooping up fish. It does not store food in this pouch

pedes have long bodies with many legs. The arachnids, which include spiders, scorpions and ticks, have bodies divided into a head region and an abdomen, and four pairs of legs.

The largest sub-group of arthropods, the insects, have bodies clearly divided into head, middle region—called the thorax—and abdomen; long feelers, good eyesight and three pairs of legs. Some, such as silverfish, are wingless, but most have wings. They include butterflies, moths, flies, mosquitoes, grasshoppers, cockroaches, bees, wasps, earwigs, true bugs, lice and termites.

The starfish, sea-urchins, brittlestars and sea-cucumbers are in a group all of their own, the echinoderms (= spiny-skinned animals).

The chordates make up the most familiar group because they include all the larger animals. Chordates have a well-developed nervous system with a main nerve-cord (the spinal cord). The main sub-group contains the vertebrates, which have an internal skeleton with a central backbone, and a brain enclosed

Above left: The blowfly is one of the most dangerous insects, because it carries dirt and germs on its feet, and then walks over food.

Above right: The giant tortoise from the remote Galápagos Islands is one of the biggest of its kind.

in a bony case. In this sub-group are fishes, amphibians, reptiles, birds and mammals.

There are two groups of fishes. Cartilaginous fishes have their skeletons made of cartilage, gristle-like material such as you can feel in your nose. This group includes sharks, rays, skates and dogfish. The other group is the bony fishes, which have skeletons of bone.

Amphibians live partly in water and partly on land. They lay their eggs in the water. The young amphibians when they hatch out spend their early life in the water and go ashore when they are adults. Amphibians include frogs, toads, newts and salamanders.

Reptiles are land animals, breathing air all through their lives once they hatch from the egg, but many of them spend a lot of time in water. They include lizards, snakes, crocodiles and alligators, turtles and tortoises.

The bodies of reptiles are covered with scales. Birds have scaly legs but the rest of the body is covered with feathers. The front legs have changed to become wings.

The mammals include all the familiar four-footed furry animals. They have hair instead of scales or feathers, and feed their young on milk produced by the mother.

Monotremes are rare, primitive egg-laying mammals, found only in Australia and New Guinea. Only two kinds are known: the duck-billed platypus and the spiny anteater. Marsupials live mostly in Australia, but some are found in New Guinea and the Americas. The mother carries her young in a pouch (the marsupium), where it can suck milk. Marsupials include kangaroos, opossums, wombats and koalas.

Plants and Animals of Oceans

The oceans cover about two-thirds of the Earth's surface. In their waters life on Earth first began, and the oceans today are the home of billions of plants and animals. As on land, the animals of the sea depend upon plants for their food, either directly, or indirectly by eating other animals which have eaten plants. Plants need light to make their own food; so they are found only in the surface waters of the oceans and in the shallow waters around the coasts.

Between the high-tide mark and the water about 30 metres deep are found the seaweeds—flat, flexible fronds covered in slime. With a few exceptions the colours of seaweeds change with the depth of water in which they are found. Green seaweeds occur in rock pools, brown seaweeds in the region between low- and high-tide marks, and red seaweeds in the deeper water. All are attached to rocks by a sticky holdfast, a root-like structure.

Most of the ocean's animals depend for their food on the plants floating in the surface waters—the phytoplankton. Phyto means plant; plankton are floating organisms living in water. Phytoplankton are tiny one-celled plants with a great variety of shapes and patterns. The diatoms have glassy, transparent shells like the base and lid of a tobacco tin, intricately sculptured in patterns of lines and dots. The dinoflagellates have even stranger shapes and are propelled through the water by whip-like hairs. Some dinoflagellates are luminescent, giving rise to a strange gleam on the sea at night.

Feeding on these plants are minute floating animals—the zooplankton. These animals include forms very similar to some of the plants. For example, radiolarians have glassy, round shells with long spines. Many baby forms of fish, shellfish, starfish, sea-urchins and crustaceans live in the plankton—floating shapes difficult to relate to their parents. Other floating animals feed on them—jellyfish, sea-slugs, shrimps and young fish.

Many whales, including the blue whales, probably the largest animals that have ever lived, feed solely on plankton, filtering the water through horny sieves in their mouths, then licking off the plankton.

Many fish hunt between the surface and the deep waters. As a protection against being eaten, fish in this region are often silvery below, so as not to show against the bright sky, and blue on top, so as not to be seen by hunters from above. The flying fish have developed a special means of escape from underwater predators, leaping clear of the water. Some hunters are very large indeed.

Sharks are big fish, while dolphins, porpoises and whales are mammals that have taken to the sea. The killer whales often hunt in packs. So do the thresher sharks, which swim in circles round a shoal of fish until they are all huddled together, and then stun them with the sharks' lashing tails.

The forms of life in the deep sea, which starts at a depth of about 460 metres, are very different. These animals live either on other animals or on the sinking dead bodies of surface-living creatures. Many deep-sea prawns and shrimps are red because this is the only colour of light that reaches these depths. Others are silver-sided or even transparent. Some have lights on their undersides, prevent-

This seaweed is in a genus (group) of brown algae known as *Laminaria.* It grows on the shores of all the world's oceans, just below highwater mark. Some people call it oarweed or oreweed.

ing them from being silhouetted against the distant sky. Hagfish swim along with their huge mouths wide open, swallowing anything.

Other hunters, called anglerfish, use a lure—a long, worm-like extension from the upper jaw, which can be waggled to look like a small animal, luring unsuspecting fish straight into the anglerfish's mouth. The deep sea is also the home of the octopus and squid, jet-propelled hunters whose own defence is to squirt out a cloud of ink to cover their tracks.

When they die, animals and plants sink to the muddy bottom, where they provide food for a rich community of animals. Molluscs and worms live in burrows, sifting the muddy water for debris. Sea-urchins scavenge along the bottom, and haddock feed on the smaller scavengers. Flatfish such as plaice and flounders live here, changing colour to match the sea bed. In the shallow water around the coasts sea-anemones cling to rocks, trapping small animals with their stinging tentacles. Crabs scuttle to and fro, and turtles paddle through the clear water.

Some creatures travel great distances during their lives. Many bottom-living fish such as cod and plaice travel to special nursery grounds to lay their eggs. The floating eggs and newly-hatched young fish live among the plankton until they have drifted to suitable places for adult life. Salmon travel from the Atlantic Ocean to inland rivers to spawn (lay eggs), while eels travel from rivers in Europe and North America to part of the North Atlantic Ocean to spawn. Even the humpback whales migrate from the Antarctic to the tropics and back every year.

A spotted eagle ray swims through the sea. It gets its name from the eagle-like appearance of its head. Eagle rays are big fishes, up to three metres across the body.

The blue whale is the largest animal. It is a mammal: the mother feeds her baby on milk.

The great white shark is one of the biggest and most dangerous of fishes.

The yellow long-nosed butterfly fish is one of many beautiful, brightly coloured fishes which live in tropical seas.

The conger eel lives in the deep oceans.

33

Plants and Animals of Europe

Europe's northernmost countries—Norway, Sweden, Finland and parts of eastern Europe, particularly Russia—are mostly blanketed by evergreen forests, with occasional patches of bog and moorland, and interrupted by mountain peaks. This dark forest region includes many familiar trees—fir, spruce, pine, larch and, in the more open parts, birch, hazel and willow.

In summer many of its animals move up to the mountain tops or north to the tundra, the treeless northern plains, to feed on plants newly exposed by the melting snow. These wandering animals include the reindeer, often herded by the Laplanders, who follow them north in summer.

The forest hides a number of large animals—the elk with their beautiful antlers, the brown bears and the wolverines, the wolves and the Arctic fox. Polecats and pine martens hunt the smaller forest creatures—young squirrels, baby birds, mice and shrews.

On the forest floor live some of the many gamebirds—grouse and capercaillie, which perform elaborate courtship dances in front of prospective mates, fanning out their tails and puffing up their chests. On the forest fringes other hunters are active—snowy owls and eagles. At the edge of the tundra live lemmings and hares. Lemmings are hamster-like animals famous for their occasional large-scale migrations when conditions become overcrowded. Thousands of lemmings set off for an unknown destination, each blindly following the ones in front, even into the sea.

Further south the evergreen trees give way to broad-leaved trees which shed their leaves in winter—oak, ash, beech, alder, aspen, lime and maple, with yellow-catkinned hazel in the

Above: Badgers are night hunters of the European woodlands. Right: The brown bear lives in the mountainous parts of the continent, especially in the Soviet Union.

Below: A gannet coming in to land. Gannets nest on the shore, and dive into the sea to catch fish.

Foxes are found in most parts of Europe. This one was photographed in the Ukrainian grasslands.

Below: A grass snake pretending to be dead. This harmless snake is found in many parts of Europe, living in bracken or near streams.

tains many trees—cork oak (the source of cork), olive, pistachios, carob, orange and lemon trees, strawberry trees, and among them vines and many fragrant herbs—thyme, rosemary, lavender and marjoram. Mistletoe is a common sight. There are several evergreens, too: holly trees, laurels, Aleppo pines, junipers and the majestic cedars of Lebanon.

On the dry hill slopes and stonier plains many of these trees remain dwarfs, stunted by lack of water. The undergrowth is a dense thicket of spiny, prickly shrubs, armoured to protect them from browsing animals. Fallow deer live here, and on the rockier slopes are wild goats and wild sheep. The bare mountains are the home of small goats called ibexes.

Geckos and lizards scuttle up and down the trees, and the bushes provide nesting sites for many warblers and other small birds. Colourful bee-eaters and rollers fly over the open slopes, and eagles soar high above them.

Large expanses of inland Europe to the east are covered in grassland. These are the steppes. The grass is stiff and feathery, and in spring many beautiful flowers emerge from underground bulbs and tubers—tulips, irises, hyacinths, buttercups and wild rhubarb. Dotted among the grasses are sage-bushes, giving off a bitter fragrance. Mole-rats and jerboas (jumping rats) make their burrows here, and marmots gorge themselves on the plants.

clearings. In this lighter forest the undergrowth is dense and tangled, rather like a cool jungle. There is plenty of shelter for the animals of the forest floor—the handsome red deer, the tiny fallow deer and the attractive roe deer. Herds of wild boars snuffle through the undergrowth for nuts and fruit. Badgers scrabble among the leaves after voles, moles, baby rabbits and bird's eggs, and squirrels scamper along the branches. Brown bears and wolves hunt here, too, and so do the rarer lynx and wildcat. Foxes and weasels are numerous, and owls glide through the trees by night.

By day the forest is alive with birdsong —the blue tits, great tits and their cousins singing from the topmost branches, chaffinches and blackbirds sounding the alarm, and woodpeckers drumming on treetrunks to attract their mates.

Around the Mediterranean Sea the climate is not so good for tree growth. The winters are wet, but the summer is long, hot and dry. Many Mediterranean plants shed their leaves in summer rather than in winter, thus avoiding the loss of too much water by evaporation from them. Others, such as holly, ivy and laurel, have leaves covered in shiny wax which waterproofs them, so they can stay on the tree all year round.

The Mediterranean forest or *maquis* con-

Plants and Animals of Asia

Southern Asia is noted for its tropical forests, in which tall trees form a closed canopy high above the ground. Rope-like plants dangle from the high branches, tree ferns rise to meet them, and the forest floor is covered in a thick layer of rotting leaves. There you can find the curious pitcher plants, which trap insects in their vase-shaped leaves, drown them in the water at the bottom, then feed on them. The parasitic plant *Rafflesia* pushes out its huge red flowers, up to a metre across, on the roots of the trees in which it lives. These flowers smell unpleasant.

Tropical Asia is also the home of the banyan tree, whose crown of branches can stretch for hundreds of metres. Giant squirrels, gibbons, lemurs and orang-utans play among the branches, watched by martens, civets and bears—sloth bears and sun bears.

The forest has many so-called flyers that actually glide—flying lizards and flying lemurs with flaps of skin stretched between their hands and feet, and flying frogs with great webbed parachute feet. There are even flying snakes, which get up speed gliding along a branch, then launch themselves into space, flattening their bodies as they do so. Hornbills, birds with huge beaks, feast on the plentiful fruit, the remains being eaten by tortoises on the forest floor. The termite-eating pangolin is one of the curiosities of the tropical forest. Armour-plated, the pangolin is a humped creature with a small narrow head, a long sticky tongue and a long scaly tail. It rolls itself into a ball when alarmed.

On the forest floor leeches and worms wriggle through the deep leaf mould, with its rich harvest of fungi. Groups of wild pigs migrate from one fruiting tree to another. There, too, you may find the rare Malayan tapir, a strange, hoofed animal with a snout like a chopped-off elephant's trunk.

The Asian rhinoceros is much more obviously armour-plated than its African relative, and the Asian leopard is often black instead of spotted, and is sometimes called a panther.

The Siberian tiger of northern Asia is one of the world's rarest animals. Only about 300 survive in the wild.

Asian cattle), Indian antelopes and water buffalo. Many kinds of deer live there, hunted by packs of wild dogs, solitary tigers, leopards, jackals and the rare Asiatic lion. Any animals which fall prey to a hunter also provide a feast for the scavengers—the hyenas, griffon vultures and adjutant storks.

Further up the mountains the forest is one of bamboos and rhododendrons. This is the home of Asia's most famous animal, the giant panda. The panda lives on bamboos like its smaller relative, the red panda, which lives higher up. Groups of wild dogs roam this forest, and the Himalayan black bear may also go there.

Above the treeline are the alpine meadows and beyond them the bleak mountain plateaus with their bitter winters. Many attractive alpine flowers grow there, such as primulas, gentians, poppies, rhododendrons and edelweiss. Wild bharal (blue sheep) and Tibetan gazelles live among the rocks, together with markhor (wild goats) and kiang (Tibetan wild asses). Musk deer graze the lush summer pastures, and snow leopards feed on the deer.

On the high plateaus are herds of yaks—bulky black animals with long black horns and thick shaggy fur, grazing on the wiry grass and small shrubs. Many yaks are now domesticated like cattle.

Surveying the scene from the air are golden eagles and vultures. Among them is the lammergeier, a vulture which carries bones to a great height, then drops them to smash them open so that it can feed on the rich bone marrow inside.

Above: The giant panda lives in the mountains of western China, feeding on bamboo shoots and small mammals, birds and fishes. Below: A banyan tree has hundreds of trunks.

Occasional caves in the forest shelter colonies of leaf-nosed and horseshoe bats, and swiftlets, whose delicate nests made of hardened saliva are used to make bird's-nest soup.

The rivers passing through the forest attract other creatures—crocodiles and river monitors (giant lizards), fish eagles, snakebirds which fish under water, tailor-birds, which sew leaves together to make their nests, turtles, macaques (monkeys) and herds of wild cattle. Here, too, the sambar, Asia's largest deer, comes to drink. Sambars and macaques must watch for enemies such as pythons, constricting snakes over 7 metres long, and the side-swiping tails of the crocodiles.

Further north is a rather different forest, many of whose trees shed their leaves in the dry season. The cobras of the Indian snakecharmers come from these forests, and so do the peacocks and their relatives the pheasants. This forest is interrupted by stretches of savannah (grasslands). In the grasslands are the Indian elephant, distinguished from its African cousin by its smaller ears, gaur (wild

Plants and Animals of North America

The northernmost part of North America is a treeless expanse of low-growing plants and reindeer moss, broken only by pools of water in spring. In winter snow covers the ground and few animals are to be found there, except some which are hibernating (in a deep sleep), the hardy musk oxen which overwinter on the high windswept plateaus, and polar bears, which spend much of their time on ice floes. In summer, when insects abound and the surrounding seas are teeming with fish and plankton, millions of animals migrate to the area from their warmer winter quarters in the forests and grasslands further south—caribou, wolves, geese, ducks and terns.

Bordering this inhospitable region is a dark evergreen forest of pines and hemlocks, merging further south into a more open forest of oaks, hickories, elms and sugar maples. Large rivers run through the open forest, and there are many meadows and clearings. Above the treeline on the mountains are grasslands rich in wild flowers. There, in the shelter of the forest fringes, the caribou spend the winter, trekking hundreds of kilometres north in spring to breed on the rich pastures which have been watered by the melting snow.

Small deer and their larger relatives, the elk and the moose, tend to stay in the forest. Even these large creatures have their enemies—the timber wolves and the mountain lions, or pumas. Wolverines, also called gluttons, are the pirates of the forest, stealing the remains of other animals' prey, and even driving away a successful hunter before it has had time to enjoy a hard-earned dinner.

North America has some large bears, the

Above: The skunk has an unpleasant smell, but rich, silky fur. Below: Bison, often misnamed buffalo, used to roam the prairies in millions until hunters killed off most of them.

black bear and the fierce grizzly, which hibernate during the bitter northern winter. Eagles hunt from the air, and the forest abounds with smaller hunters—red foxes, bobcats, hawks, owls and skunks. These animals catch and eat voles, rats, shrews, small birds, squirrels and wild turkeys.

Tree-dwellers are many and varied. There are grey squirrels, and flying squirrels which use flaps of skin stretched between their hands and feet as parachutes for gliding from tree to tree. Chipmunks are like small, brown, striped squirrels, and they store nuts to help them through the cold winter. Porcupines also climb trees, stripping them of bark as they go. The opossum, North America's only marsupial (see pages 44-45), lives in the trees.

The rivers attract raccoons, which have faces like masked bandits and bushy, ringed tails. They feed on water animals such as fish, frogs, crayfish and mussels. Larger creatures—the Alaskan brown bears—also fish in these rivers for salmon. The most famous water creatures are the beavers. They fell trees to build dams and so make ponds in which to build their homes, which have underwater entrances for safety. The logs are floated down to the dam, often along canals

quills and their ability to burrow rapidly. Pocket gophers escape to ready-made underground homes. So do the prairie dogs, hamster-like short-tailed ground squirrels, whose colonies are so enormous that they are called prairie dog 'towns'. One of the largest towns covered an area almost as big as Sri Lanka and contained 400 million prairie dogs, in family burrows.

The prairie hunters range from the badgers and kit-foxes to wolves, coyotes and occasional mountain lions. Less easily-seen attackers are rattlesnakes, which can detect prey at night by the warmth coming from their bodies. Parts of North America are deserts, where only cacti and a few wiry plants grow. Even there a few small animals manage to survive.

The coasts of North America are varied and rich in wildlife. The Arctic waters off the northern coasts provide food for seals, walruses and many seabirds.

Along the western coastal strip are stands of redwoods (sequoias), massive evergreen giants which include some of the oldest and tallest trees in the world. Sealions live along the nearby shores. Further south and east, in the warm waters of the Gulf of Mexico, are swamps containing mangroves, strange trees whose roots protrude above the water like elbows. Alligators lie in the mud below them, under trailing tangles of Spanish moss dangling from the branches.

dug by the beavers themselves. The beaver lodge is made of sticks cemented together with mud. The beavers use underwater rooms to store tree bark as a winter food supply.

The mountains have their own special wildlife. Mountain goats and wild bighorn sheep clamber among the rocks. Below ground are the pocket gophers, living on bulbs and roots. The name 'gopher' comes from a French word meaning 'honeycomb'—a good description of the many tunnels made by these tiny rodents. Other animals with not such an all-year-round food supply have to make special arrangements for winter. The marmots store fat in their bodies, then hibernate, while the pikas (creatures rather like guinea pigs), collect hay, drying it in the sun on a rock and building small haystacks for winter use.

South of the great forests are the prairies, wide expanses of flat grassland. There is little shelter on the prairies, and the only hope of escape from predators is to run away or burrow in the ground. The large herds of pronghorn antelopes and deer have a system of sentries posted to warn of approaching enemies. The biggest prairie animal is the bison, or American buffalo.

Porcupines have two defences—their spiky

The puma is a large member of the cat family. It hunts mostly at night, feeding on deer and many smaller animals. It is also known by the names of cougar or mountain lion.

The sugar maple is one of the most beautiful trees of North America. In fall its leaves turn a bright, glowing red. These trees are in New England.

Plants and Animals of South America

South America is one of the most mysterious continents. Large areas of it remain unexplored, and probably contain plants and animals unknown to man. Down the west side are the high peaks of the Andes, their cold sides covered with a few scanty plants below the permanent snowline. A little further down is a band of 'elfin woodland'—scattered dwarf trees—which merges into the cool cloud forest, a region that is almost always enveloped in cloud, with a glistening dew decorating every plant.

The air in the cloud forest is so moist that many plants can manage without water from the ground. Called epiphytes, they grow on the branches and trunks of trees, absorbing water from the air through dangling roots. Epiphytes include mosses and ferns, many beautiful orchids, and the large, succulent, spiky plants called bromeliads, which are grown elsewhere as indoor plants. In the wild they have pools of water in their hollow centres in which small frogs, tadpoles, salamanders and other tiny water creatures make their homes.

Between the high peaks are high plateaus of grazing land, where some of South America's most characteristic animals are found. There are four relatives of the camel there. The

Above: The toucan with its huge bill which is actually very light.
Below: The iguana is a large lizard, about two metres long. These two animals live in the tropical forests.

domesticated llamas are bred for meat and as packhorses; and the alpacas are used for wool and meat, rather as sheep are reared elsewhere. The wild vicuñas live as high as vegetation can be found, and the guanacos roam in herds from the high Andes to Patagonia in the far south.

The mountain plateaus are also the home of the chinchillas, small squirrel-like animals famous for their soft fur. They live in cracks in the rocks, unlike their relatives on the grassland plains, the viscachas, which live in vast underground burrows. Pumas (mountain lions) roam the plateaus, and so do smaller predators such as weasels and skunks. The Andean condor, the world's largest flying bird, lives on carrion (dead animals).

In the southern part of the continent are large stretches of grassland, the pampas. Unlike the African savannahs and the American prairies, the pampas have no large herds of grazing animals. Small groups of guanacos graze there, and rheas, large flightless birds, trot around the plains, their long necks helping them to see above the tall grasses. Many rodents such as viscachas, agoutis and pacas live in the grass, eating the plentiful seeds. Skunks and pampas foxes prey on rodents, and in turn are preyed on by pumas and jaguars, large members of the cat family.

The pampas are the home of anteaters and armadillos. The armadillos are well-named, as they are covered with large horny plates. When frightened they can burrow extremely fast, but if there is no time for that, they roll themselves up into an armour-plated ball, like a millipede or woodlouse. The giant anteater lives on a diet of ants and termites. It has grey

Sloths spend their lives hanging upside-down from branches. This one is a two-toed sloth, and there is a three-toed species as well.

Top right: A typical rain forest of South America, with quick-growing trumpet trees *(Cecropia).*

The three-toed anteater or tamandua spends most of its time in the branches of trees.

fur and a magnificent bushy tail. It has long claws for breaking into termites' nests, and a long sticky tongue to pick up its prey.

The Amazon jungle, in the widest part of South America, is a great river-basin of tropical forest, flooded frequently by the huge Amazon River and its tributaries which flow down from the Andes. The forest is a gloomy place with tall trees whose branches form a canopy over the forest. Below this are tree ferns and shade-loving shrubs, and a thick carpet of leaves. Many of the tall trees have buttress roots—wing-shaped above-ground props, whose folds provide shelter for many forest animals. Lianas, long rope-like plants, dangle in tangles from the tree-tops to the forest floor, borrowing support from other plants in their struggle to reach the light. Climbers and creepers are everywhere.

In the high branches live many kinds of monkeys: agile spider monkeys, noisy howler monkeys, the appealing marmosets, tiny douroucoulis, and capuchins. The monkeys of the Americas differ from their relatives elsewhere in having prehensile (grasping) tails. They use these tails as a fifth limb, curling them round branches.

Sloths are large ape-like creatures with small heads and thick, matted fur. They hang upside-down from branches, moving incredibly slowly. Fruit and leaves are so plentiful that the sloths do not need to do much work. Their coats are covered in green algae (tiny plants), which help to camouflage them.

Dragon-like iguanas, which are huge lizards, lie along the branches, simply falling from one bough to another if they want to get to the ground. Fierce predators include snakes—boa constrictors, fer-de-lance and

bushmasters—and swarms of army ants which eat anything in their path.

There are no seasons in the tropics, and flowers and fruit can be found all year round. These attract nectar-feeders—gaily-coloured humming-birds, shining morpho butterflies and fruit-eating birds such as toucans, which have enormous brightly-coloured bills. Gaudy macaws (parrots) feed on the flower seeds.

The rivers are teeming with life, including many of the colourful fishes you see in aquariums, such as tetras, cichlids and angelfish. There are also piranhas, aggressive fishes that move in huge shoals, tearing the meat off the bones of any animal that floats their way. Caimans (South American alligators) lie in the shallows. At the water's edge you can see capybaras, the world's largest rodents, and giant river otters. Anacondas, water snakes 10 metres long, squeeze animals to death.

Plants and Animals of Africa

Africa is a continent of contrasts. It also has some of the most exciting wildlife in the world, though hunting is gradually reducing the numbers of wild animals. In the north the land bordering the Mediterranean Sea is similar to the Mediterranean lands of Europe (see pages 34-35). Then comes the vast Sahara desert. The plants and animals of this and other desert regions are described on pages 46-47. Most of the rest is either savannah (grasslands) or forest.

The main forest regions are in central Africa and in the west just south of the Sahara. This area has a great deal of rain all the year round. Most of the forest is made up of tall trees whose high branches form a cover that cuts off most of the light from the Sun. Near rivers and well-used paths there is more light and the forest becomes thick jungle, with many tangled, low-growing trees and bushes.

Many kinds of monkeys live in the forests. They include the chimpanzee, the animal most like Man. Larger animals include buffaloes, some small antelope and the okapi. This shy creature is the only relative of the giraffe, but it does not have such a long neck. It hides deep in the forest, and is so hard to see that nobody knew it existed until 1900. There are also some elephants and a small hippopotamus, called the pigmy hippopotamus. The only large hunters in the forest are leopards.

There are a great many birds and insects in

The African elephant is the largest land animal. It is almost as tall as the height of two men. Elephants roam in herds, eating grass and trees.

Aloe trees and dry, wiry shrubs and grasses form the landscape in this scene in South-West Africa (Namibia).

the forests. Sunbirds feed on the nectar of tropical flowers, which bloom all the year round. Brightly-coloured parrots, hornbills and turacos, like the monkeys, are very noisy. Brilliant butterflies skim through the trees, while on the ground columns of driver ants march through the leaf-litter, eating any creature that does not get out of their way. There are myriads of termites, which build huge, rock-hard mounds as their nests.

The savannah is a mixture of grass and trees. It is called the veld in South Africa. Some parts of the savannah are dry and inclined to be dusty, others are wet and green. Many of the grasses grow very tall. Among the trees the most curious is the baobab, which has a barrel-shaped spongy trunk in which it stores water, and a crown of branches which provides a home for birds.

The grassland becomes so dry during the hot season that lightning sometimes starts fires. Fortunately the grasses quickly regrow. Animals either run away from the flames or hide in their burrows underground.

In such open country, where food may be

scarce during the dry season, animals that can travel quickly do best, both in looking for food and escaping from predators. Herds of elephants, buffaloes, zebras and antelopes of several species are continually roaming the plains. Because the various types of animals like different kinds of food a number of herds can graze in the same area at the same time. Some eat the leaves and shoots of trees. Giraffes, having long necks, can eat trees much more easily than they can bend down to have a drink from a pool. The gerenuk, a little antelope, stands up on its hind legs to reach the tops of shrubs. Elephants feed at all levels, pulling up grass or pulling down branches to eat.

Several hunting animals roam the grasslands, preying on these large herds. Lions hunt in small groups, leopards rely on quiet ambush, while the smaller cheetah can outrun any savannah grazer in a quick sprint. There are also packs of hunting dogs and hyenas. Hyenas get part of their food by scavenging, eating dead animals or the leftovers from the dinner of lions and other hunters. When the

Above left: The cheetah is one of the big hunting cats of Africa. It can run faster than any other animal. Above right: The giraffe's long neck enables it to browse on the leaves of trees.

hunters and the hyenas have finished, vultures pick the bones clean.

On the eastern side of Africa is the Great Rift Valley, a fault in the Earth's crust of which the Red Sea and the Dead Sea are also part. The Rift Valley is deep and wide and contains huge lakes. Vast colonies of colourful pink flamingoes nest by the lakes, trailing their beaks in the water to filter out food. Pelicans are a common sight, sometimes hunting in flocks to drive fish into shallow water where they can easily be scooped up.

Also competing for the fish are grebes, cormorants, darters and kingfishers. Jaçanas, long-legged birds about the size of moorhens, run across the floating leaves of water-lilies, their long toes spread wide to prevent the birds from sinking.

In the rivers and lakes there live crocodiles, huge reptiles which eat any animal unwary enough to come within reach of their long jaws. Some of the largest African animals, the hippopotamuses, also live in lakes, keeping cool in the water by day and coming to land at night to browse.

Plants and Animals of Australia

Australia is one of the world's driest continents. More than half the country—the 'Red Centre'—is desert, surrounded by a semicircle of forest and savannah (grassland). The far north of the continent is tropical rain forest, but elsewhere the most common forest consists of gum trees.

Gums are eucalyptus trees, with pale waxy leaves and sticky resin-coated trunks and twigs. Some eucalypts reach heights of more than 90 metres. The eucalyptus oil they contain gives off a vapour which causes the forests to be enveloped in a blue haze in the sunshine. This haze gave the Blue Mountains of New South Wales their name. Eucalyptus trees do not cast too deep a shade, so there is a rich undergrowth of shrubs providing shelter for many wild creatures. Towering above the shrubs are giant tree ferns, with stems like fat palm trees, and crowns of fern fronds at the top of the stems.

The mammals of Australia are mostly marsupials, a group of animals found only in Australia, New Guinea and the Americas. Marsupials carry their young around in pouches on the mother's belly. As soon as it is born, the undeveloped baby marsupial find its own way into the pouch, which forms its home for several months. Even after it no longer needs its mother's milk, it returns to the pouch for shelter. A female kangaroo has been seen fighting with a 23 kg youngster still wanting to travel in her pouch!

Kangaroos and their smaller cousins the wallabies are the best-known marsupials. They travel by leaping with their powerful hind legs, dangling their short front paws like

The golden acacia, known in Australia as the wattle, is the country's national flower.

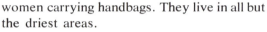

A swimming platypus rootles head-down in the river bed for its food.

women carrying handbags. They live in all but the driest areas.

Koalas, which look like animated teddy-bears, spend their time in gum trees and feed on the leaves. Other tree-dwellers are the possums, or phalangers, small furry animals. Some possums glide, using flaps of skin stretched between their hands and ankles. The sugar glider and the honey possum live mainly on a diet of nectar from flowers, as do some big birds, the honeyeaters. Many of Australia's flowers, such as the banksias and eucalypts, are large and sturdy, which enables them to cope with hefty visitors; they are coloured red or orange, and attract the birds.

Another group of mammals found only in Australia and New Guinea is the monotremes, mammals which lay eggs, but later keep their young in a primitive pouch and feed them on milk. There are only two known monotremes. The duck-billed platypus is a beaver-like animal, with a flattened tail, webbed feet and a large flattened duck-like bill. It lives in rivers and ponds, making a burrow in the bank.

The spiny anteater or echidna is rather like a porcupine, and can roll up into a ball if disturbed. If it is alarmed while on sandy or soft ground it burrows rapidly, leaving just its spiny back showing. It has a long, sensitive muzzle and a sticky tongue for scooping up the ants on which it feeds.

Australia has some fascinating birds. The male lyre bird has ornate plumage which attracts females, and can raise his feathers to form the shape of an exquisitely patterned

lyre, an ancient musical instrument. The bower bird attracts his female by building an elaborate home decorated with shells, flowers and leaves.

Some Australian birds cannot fly. The cassowaries and emus, up to 2 metres tall, live mainly in the grasslands which stretch inland from the forests to the edge of the desert. Emus can run at speeds of up to 48 kph.

The grasslands also provide cover for dingoes. Dingoes are wild dogs introduced to the continent thousands of years ago by the Aborigines, the first people of Australia. Dingoes are successful hunters, living in packs. Extremely intelligent animals, they work together to trap even the largest kangaroos. Less aggressive are the vegetarian wombats, similar to badgers, living in extensive burrows below ground.

In Australia's desert heartland most life is found around the few springs and the temporary pools and rivers that form after rare rainstorms. There are flocks of brightly-coloured cockatoos and budgerigars (also known as parakeets), fish and other water creatures, and a great variety of snakes and lizards. The sand goanna is a long lizard which rears up on its hind legs to threaten an attacker. The bearded dragon, also a lizard, can stick out a bristling frill around its mouth, and the frilled lizard can open its mouth wide, raising a huge ruff all around its face.

Above: Koalas live in eucalyptus trees, feeding on their leaves. Above right: The lyre bird gets its name from the shape of its tail, which resembles an old musical instrument. Right: A mother kangaroo carries her baby, called a joey, in her pouch.

Plants and Animals of Deserts

To most of you the word 'desert' probably conjures up a picture of a barren land having no life except for a few spiny cacti and wiry shrubs. Visiting a desert in the heat of the noonday sun might well give anyone this impression—few creatures have learnt to cope with the intense dry heat of the midday desert, but at dawn or dusk, and during the cold desert night, life in abundance can be found.

One of the biggest problems of living in a desert is not simply the small amount of rain that falls every year, but the irregular times at which it falls. Some deserts have rain only once every 12 years or so and many receive rain only in occasional heavy downpours.

The desert plants cope with the water problem in a variety of ways. The tiny desert annuals, which provide many of the brilliant carpets of flowers that appear after rain, germinate, grow, flower and produce seed in less than six weeks. Their seeds have a coating which is not penetrated by a light shower, but heavy rain soaks through and makes the seeds sprout. Their seeds can wait many years for enough rain to start growth.

Other plants shed their leaves and even

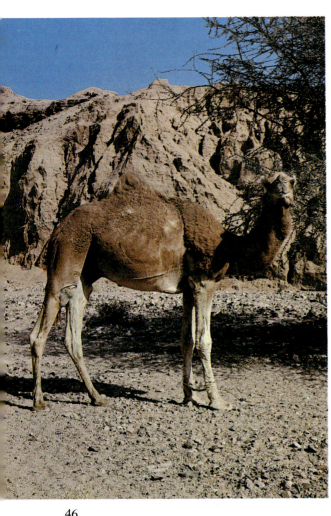

Above: Scattered, wiry plants are all that can grow in the shifting sand dunes of White Sands National Monument, New Mexico. Right: Bright flowers bloom after rain in a desert in Peru.

Left: The Arabian camel has been called 'the ship of the desert' because it can stand great heat and go for long periods without food or water. This kind of camel has one hump; the camel with two humps is the Bactrian camel of Asia.

their branches at the start of the dry season, passing the dry periods underground, full of stored food and water—a choice find for any animal that digs them up.

Cacti store water in their tissues, forming a valuable source of water for the desert animals. Some birds, such as the elf owl of Arizona, use giant cacti as nest-sites, pecking away to make a suitable hole. Many desert

plants have spines which protect them against hungry and thirsty animals, and some plants in Middle Eastern deserts are poisonous.

Some desert plants, such as the Australian river red gums and the American mesquites, have a double root system—long roots which reach permanent underground water sources, and extensive shallow roots which catch as much rain as possible before it evaporates. Other plants, such as creosote bushes, have only shallow roots, and grow some distance apart, showing that each plant needs a minimum area to be sure of getting enough water. A few plants, such as the brittle-bush, produce poisons which prevent others growing near them and 'poaching' their water supply. The resurrection plant can dry out completely and be blown around the desert like a bundle of dead twigs. As soon as rain falls it becomes green.

Some desert animals avoid the dry season by hibernating. During this deep sleep their body activities slow almost to a standstill, saving valuable energy and food use. Spadefoot toads dig holes in the sand and line them with a sticky substance that prevents them from drying out.

Kangaroo rats and jerboas spend the heat of the day in cool underground burrows, plugged to keep moisture in. Birds, such as desert swifts and the North American poor-will, sleep at night, saving their energy for flying around in the hot desert day.

Kangaroo rats store seeds in their burrows

Above left: Succulent plants which can store water, growing in a desert in South America. Above right: The jerboa is a tiny animal, related to rats and mice, which lives in the deserts of North Africa.

to eat in the dry periods or during the hot daytime, and so do the harvester ants.

Like the seeds of plants, many insects and small water creatures such as shrimps and desert fish pass dry periods as tough, drought-resistant eggs or cocoons. They swiftly hatch with the first heavy rains. Desert frogs can complete their whole life-cycle, from eggs to adults, in a month.

Some animals can also store water. Tortoises do so in sacs under their shells, while some desert frogs and toads become swollen with water before burying themselves in the sand. Camels store water in their humps in the form of fat, which releases the water when acted on by the body juices. Fat-tail desert mice and sheep use a similar process.

Heat presents a problem for many desert creatures. Reptiles cannot tolerate great heat, and those which do emerge in the day tend to stay in the shade. Most desert animals are mainly active at night, when the desert is very cool and surprisingly humid—some deserts have as much as 250 mm of dew in a year.

Certain kinds of animals are found in almost all deserts. Each has its jumping rats: kangaroo rats in America, jerboas and gerbils in Africa and Asia and marsupial kangaroo mice and pitchi-pitchis in Australia. The tiny 50 mm-high kangaroo rats can jump 450 mm off the ground. Large flightless birds are also found in deserts: ostriches in Africa, rheas in South America, and emus and cassowaries in the Australian outback.

Plants and Animals of the Poles

The polar regions are some of the strangest places on Earth. In summer the Sun shines almost all night as well as all day. The winter is dark, and the Sun barely rises above the horizon. The poles are really deserts of ice and snow. They are bitterly cold for much of the year; for only a few weeks in the north does the temperature rise high enough for plants to grow.

The North and South Poles are very different places. The South Pole lies on the vast continent of Antarctica, which is permanently covered by a great domed ice cap, with glacier tongues creeping into the sea. Blizzards sweep down the ice cap for much of the year. Antarctic life occurs only around the shores, where the snow melts for a short time each year, and on the sea ice, where birds can fish in the cold waters. The land animals of the Antarctic are penguins, flightless birds more at home in the sea than on shore, and a few insects. The main life is in the sea around Antarctica—whales and fishes.

The North Pole, by contrast, lies in the Arctic Ocean, most of which is frozen for part of the year. The ocean is surrounded by groups of islands and the northern parts of Europe, Asia and North America. Because the sea does not cool down as readily as the land, the Arctic is warmer than Antarctica and supports much more life.

The soil on land around the Arctic Ocean is shallow and not very fertile. Only a thin top

Polar bears are equally at home on the ice and in the bitterly-cold waters of the Arctic Ocean.

Reindeer and their North American cousins the caribou migrate north in the brief Arctic summer, going south again to find food in winter.

layer ever thaws out. Below is a permanently frozen layer, the permafrost. However, many plants grow in the region between the permanent snow and the tree-line, the latitude at which it is warm enough for trees to develop. This region is called the tundra. The most successful plants here are lichens, strange crusty growths which are made up of a fungus body with food-producing algae inside.

Where the snow melts in spring and forms pools and boggy ground, mosses flourish. Their tiny rootlets need little soil. The flowering plants are mostly dwarf cushion-like plants. By keeping close to the ground they get as much heat as possible from it in summer, and escape the worst of the winter winds. Most of them store food in their underground parts, and so make the most of the short spring growth season. During this brief period the tundra is suddenly covered with carpets of flowers.

All the animals which pass the whole year in these regions are warm-blooded, and they have special systems for keeping up a high body temperature despite the cold outside. Many of them develop extra coats in the winter. The musk ox grows an inner mat of fine, long fur every autumn. This provides such good insulation that the snow does not melt where an ox lies down.

Polar bears, seals and walruses have thick layers of fat beneath the skin for insulation. This also makes them more buoyant when swimming. Arctic foxes have small ears so they do not lose too much heat from them, but many polar animals save heat by managing with very cold feet—warm body surfaces lose a lot of heat to the cold air outside. The leg temperatures of reindeer and husky dogs can be 28°C lower than those of their bodies.

Many insects avoid the cold polar winters

A baby penguin shelters between its parent's feet in the bleak wastelands of Antarctica.

altogether by spending them in the form of chrysalids, immature forms inside hard cases. The spring ponds provide good homes for insect life, and mosquitoes are the curse of the polar traveller. Some animals, such as lemmings, store food in underground burrows during the summer, and stay below ground for most of the winter, snug under an insulating layer of snow. Other animals hibernate in warm holes. Hibernating animals cool right down, and all their body processes slow almost to a standstill while they sleep. They use very little energy, living off fat formed during the previous summer. Many insects and small water creatures use a similar method of escape from the winter.

Another means of escape is by migration. Many birds—ducks, geese, plovers, cranes and terns—breed in the tundra in summer, feasting on the abundant insects. With the approach of winter they fly south. The Arctic tern prefers to live constantly in polar summers, flying from the Arctic to the Antarctic and back every year. Vast herds of caribou and reindeer migrate from the northern forests to the tundra every spring, followed by wolves and other predators. Animals that prey on smaller creatures—Arctic foxes, weasels and snowy owls, for example—also hunt further south in winter and move north in spring.

Many animals, hunters and hunted, change colour with the seasons and are therefore not so easy to see. Stoats, Arctic hares and ptarmigans (large game birds) are brown in the woodlands where they live in summer, but turn white when the snow comes.

49

Living World Inquiry Desk

Why is the peacock so brightly coloured and the peahen so dull?
The peacock's bright colours serve two purposes. His brilliant plumage attracts a mate in the breeding season. More importantly, it also lures enemies away from the nesting peahen. The peahen is dull-coloured, and so she can blend into her surroundings while she is nesting. This is known as protective coloration, and it helps to keep her hidden from her enemies. Nearly all female birds are less brightly-coloured than males, so their young have a better chance of survival.

? ? ?

Why do some animals hibernate?
The word 'hibernation' means winter sleep. Some small animals such as dormice, hedgehogs and skunks hibernate to stay alive during the cold winter months, when their food supply disappears. During the late summer and autumn such an animal eats lots of food which is stored as fat in the body. As winter gets nearer, the animal finds a sheltered place such as a hole or burrow and gradually falls asleep. During hibernation the animal's breathing and heartbeats slow right down. Its body temperature drops until it is only slightly warmer than the surrounding air. In this state the animal's body burns up energy very slowly, and it can last the whole winter on the reserves of fat in its body. When it wakes up it is very thin and hungry.

Once spring comes, hunger and the rise in temperature wake the animal up once more. Many cold-blooded animals such as frogs and carp also hibernate, as do some insects including bees and various butterflies.

? ? ?

How do camels manage without food and water in the desert?
Camels can travel for several days without food and water. For this reason they have been used as beasts of burden in the deserts of Africa and Asia for thousands of years. Food, as fat, is stored in the camel's hump, and water in its stomach. Before starting a journey the camel is forced to eat and drink as much as possible. As the camel eats, its hump is filled with fat tissues causing it to rise up firmly from the camel's back. It can survive for several days on this store of fat. When the hump is empty it loses its firmness and becomes floppy. A camel has three stomachs. Two of these contain muscular pockets in which water can be stored. When the camel needs water the muscles relax, letting the water out.

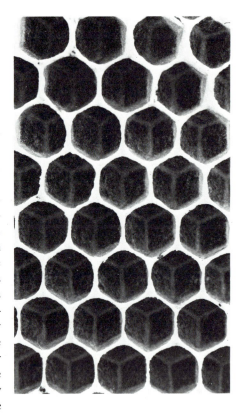

The six-sided cells which form a honeycomb.

? ? ?

How do bees make honey?
Honey is made from the nectar of flowers. Each worker bee collects nectar with its tongue and carries it back to the hive in its honeysac—a bag-like sac just in front of the bee's stomach. While the nectar is still inside the honeysac enzymes (special chemicals) convert it to the various sugars that make up honey.

The bees empty this sugary fluid into honeycombs which are then capped with wax. Before sealing the combs the bees dry out the honey by creating a strong draught through the hive. They do this by gathering at the entrance to the hive in a swarm and fanning air into the hive while they flap their wings together.

? ? ?

Can spiders live under water?
There is one species of spider that does this: *Argyroneta aquatica*, the water-spider, which is found in Asia and parts of Europe. Although it can live on land, it is the only spider that spends most of its life under water. It even rears its young under water.

In the spring the female builds a bell-like waterproof web between the submerged stems of pond plants, just a few centimetres below the surface of the water. In order to provide a supply of air the water spider rises to the surface, traps a bubble of air and carries it into the web. She does this several times until she has a collection of air bubbles about the size of a hazelnut. In this way her web becomes a miniature diving bell, in which she lives until the autumn.

During the mating season the male spider builds a similar bell near to hers. The two webs are linked by a silken tunnel. When the young spiders hatch they live in the web until they are old enough to build their own diving bells and collect their own air supply.

? ? ?

What is a coelacanth, and why do people call it a living fossil?
A coelacanth is a large fish, nearly two metres long. Scientists found many fossils of these fishes in rocks, and were able to work out that they first came into existence about 350,000,000 years ago, and had apparently died out about 60,000,000 years ago.

To everybody's astonishment, a boat trawling for fish off the coast of South Africa in 1938 hauled up a live coelacanth. The fish was almost identical to the ones known as fossils. Several more coelacanths have been caught since off the Comoro Islands, which lie north-west of Madagascar.

Very few species of animals survive so long without changing, and that is why those that do are called 'living fossils'. Some insects, particularly dragonflies, have changed very little over millions of years.

? ? ?

Why do leaves change colour in the autumn?
Leaves are green in summer because of the presence of chlorophyll, a

Above: A water-spider prepares to weave its underwater home. Below: The white fur of a polar bear does not stand out against snow and ice.

green pigment found in nearly all plants. Chlorophyll converts sunlight into chemical energy as part of the plant's food-making process. There are other pigments in leaves, such as xanthophyll which is yellow and carotin which is orange. During the summer these colours cannot be seen, but during the autumn and winter the chlorophyll disappears as the food-making process comes to an end. This leaves the other pigments behind and the leaves take on their autumn colours before dying.

? ? ?

Why do animals in snowy countries have white coats?
Animals are safer when they blend into their surroundings. Animals such

as the polar bear and the snowy owl are white, so they match the snow and ice around them. As a result it is difficult for their hunters or for their prey to notice them.

This sort of disguise or camouflage is very common in the animal world, and takes many different forms. Some animals, such as tigers and toads, are coloured or patterned in the same way as their surroundings. Others, such as prawns and chameleons, even change colour to match their background.

Some animals, particularly insects, actually resemble other objects. When moths fold their wings they blend into the bark of trees; stick insects look exactly like the plants on which they feed, and some butterflies at rest look just like leaves.

The Human Body

Human beings are the most highly developed mammals on Earth. With their superior brains they have conquered nature and can survive almost anywhere in the world. They have tamed other animals for their own use. They have harnessed the power of natural resources to provide such things as electricity. They have the ability to construct things.

Despite all this, the physical structure of Man—that is, the human species—is much the same as that of other mammals. He also has bones, nervous system, muscles and organs.

There are more than 200 bones in the human skeleton. These form a framework which protects and supports the body. The spine—a chain of 26 interlocking bones—is the body's central column. It links the arms, legs and skull. Where bones meet they are linked by joints. Some joints are rigid and cannot move. Those that can move are covered by plates of cartilage (gristle) which glide smoothly over one another for ease of movement. Over the bones are muscles. These are composed of bundles of fleshy fibres. They are the motors of the body, enabling Man to make quick, intricate movements.

Human beings are omnivores: they eat both

The human skeleton, showing some of the more than 200 bones it contains.

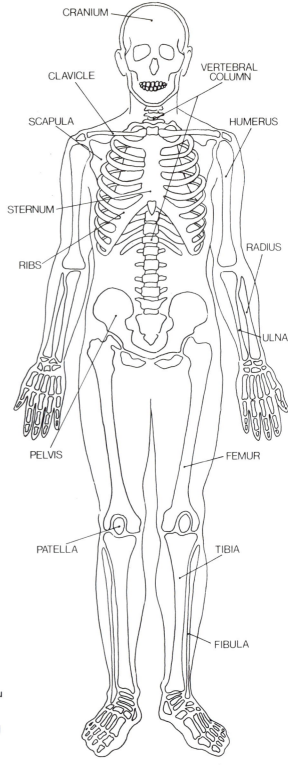

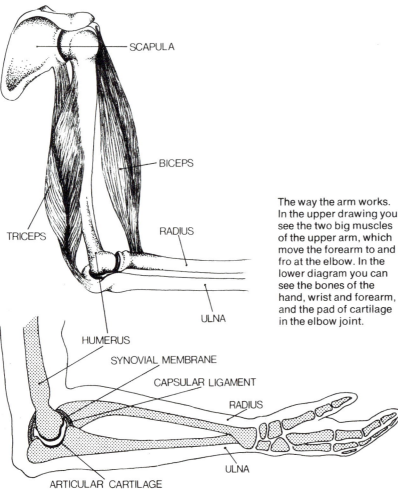

The way the arm works. In the upper drawing you see the two big muscles of the upper arm, which move the forearm to and fro at the elbow. In the lower diagram you can see the bones of the hand, wrist and forearm, and the pad of cartilage in the elbow joint.

plants and animals. The digestive system breaks down these complex foods into simple proteins, fats and sugars. These are then absorbed into the bloodstream as body-building and energy-giving substances.

About 5 litres of blood circulate continuously around the body, pumped by the heart. The heart beats at a rate of about 70 beats a minute. With each beat oxygen-rich blood from the lungs enters the left side of the heart to be circulated to the organs and tissues,

52

while stale blood is pumped from the right side for circulation to the lungs. There it absorbs more oxygen.

Most mammals have a covering of fur,

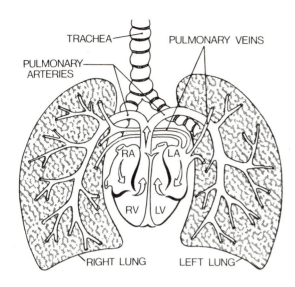

TRACHEA PULMONARY VEINS
PULMONARY ARTERIES
RA LA
RV LV
RIGHT LUNG LEFT LUNG

which protects them from the cold. Man, by comparison, has little hair. Instead he has an insulating layer of fat under the skin. The small hairs covering the body react to cold by standing upright and so trapping warm air in contact with the body. In a warm temperature the sweat glands produce a salty liquid which evaporates and cools the surface of the skin.

The nervous system is the messenger of the

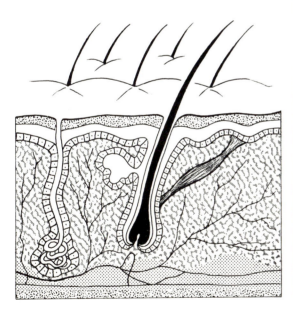

body. It includes the brain, spinal cord and a network of nerves and nerve fibres all over the body. Each nerve carries particular information to the brain. Nerves in the eye, for example, carry information about light, enabling us to see.

The human brain in comparison to body size is enormous, much bigger than that of other animals. Man is far above the level of intelligence of other mammals. He can reason,

Right: Man is closely related to the gorilla, but his brain is bigger.

Left: The working of the heart and lungs. Air containing oxygen enters the lungs through the trachaea, and the oxygen is absorbed into the blood. The left side of the heart pumps blood from the lungs to the rest of the body; the right side pumps blood from the body to the lungs. The letters RA, LA, RV and LV stand for right and left atrium and right and left ventricle.

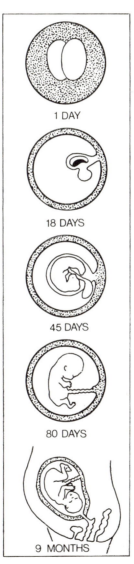

1 DAY

18 DAYS

45 DAYS

80 DAYS

9 MONTHS

Above, left: A section through the skin showing how the hairs are attached to it.
Above: The development of a baby from the first day, when the first fertilised cell divides, to nine months when the baby is ready to be born.

solve problems and put plans into operation.

Humans must reproduce in order for the species to survive. Like other female mam-

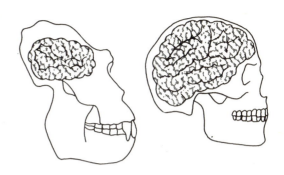

mals, women carry their developing young inside them in a womb and suckle them after birth. A male and female cell are needed for a human child to be created. Female cells or eggs are stored inside a woman in two sacs known as ovaries. These lie on either side of the womb and are attached to it by the fallopian tubes. Male cells or sperm are stored in two testes; these are contained in a small sac hanging below the penis.

Once a month one egg travels down the fallopian tube towards the womb. For it to be fertilised it must meet up with a male cell. This travels from the man's penis into the woman's vagina and into the fallopian tube. If it reaches the egg fertilisation takes place. If the female egg is not fertilised it passes out of the woman's body with the lining of the womb as the monthly period or menstruation.

The fertile egg plants into the mother's womb and starts to grow. By the time it is six weeks old the foetus, as it is known, has a large head and tiny fingers and legs and is already beginning to look human. The foetus stays in the womb for nine months. It is protected in a fluid-filled sac and is fed from the mother's bloodstream through a special organ called the placenta. The blood flows through a lifeline connecting the baby and the placenta, called the umbilical cord.

After about nine months the foetus is fully developed and the baby is born. The doctor or nurse attending the mother cuts the umbilical cord close to the baby's abdomen. This cord later falls off, leaving a permanent scar known as the umbilicus or navel.

The young of some animals can look after themselves from a very early age, and many baby animals—horses and giraffes, for example—get to their feet within minutes of being born. Human children, however, need the care and protection of their parents for many years before they are old enough to care for themselves.

The Descent of Man

You may often have wondered 'Where did Mankind come from, and why are people so different from animals in some ways, and so like them in others?' People have been asking questions like these for thousands of years. The answer is not easy to piece together from the evidence we have.

Written accounts of historical Man and his activities only go back about 5,000 years, because people had not learned to write until about 3000 BC; but we know a great deal about Man's earlier story because we have found and dug up remains of his houses, the things he made, and the skeletons of ancient people. Archaeologists (people who study the remains of ancient human life and activities) and anthropologists (people who study Man in a scientific way) have combined to piece together the story of the descent of Man from ancestors which were probably not man-like at all to the present day.

Studies of the bodies of people and animals show that Man belongs to a group of mammals (animals which suckle their young) called Primates (those which come first). The other Primates include apes, monkeys and lemurs. The apes, sometimes called anthropoid apes from a word meaning 'manlike', are more closely related to Man than they are to some other monkeys. There are four anthropoid apes: the gorilla, orang-utan, gibbon and chimpanzee.

However, it would be wrong to say, as some people do, that Man is descended from monkeys. It would be more correct to say that modern monkeys and apes and present-day Man are descended from a common ancestor, a kind of mammal that has long since died out. This descent has taken place through the process known as evolution.

The theory of evolution was first stated in the 1850s by two great scientists, Charles Darwin and Alfred Russel Wallace. By studying animals of all kinds, they came to the conclusion that only those species of animals best fitted to survive actually did so. Darwin called this process 'natural selection'. You are like your parents, but slightly different from them. These slight differences may, over a long period, make all the difference between whether you and your descendants are adjusted to the conditions in which you live, and so survive or perish.

A modern example of natural selection in progress is shown by the peppered moth. The older type of moth has a light greyish, mottled colour, and when it sits on the bark of trees with similar colouring it is very difficult for a hungry bird to see it. In this way many of

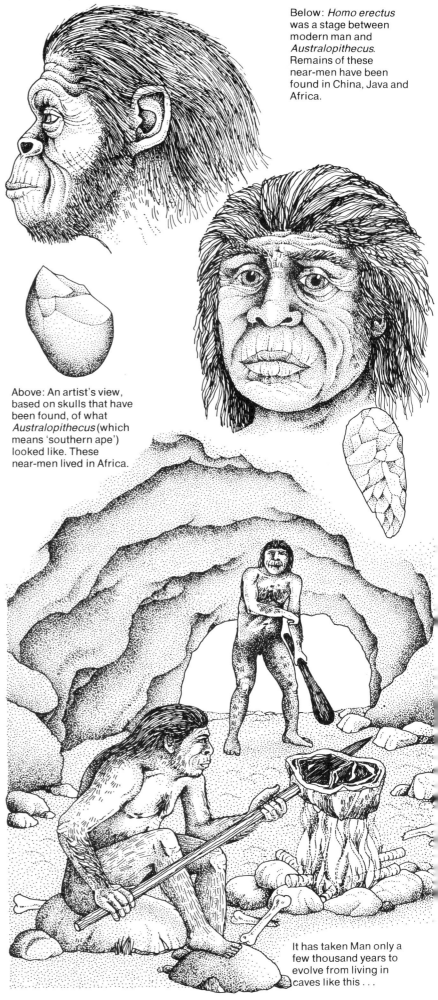

Below: *Homo erectus* was a stage between modern man and *Australopithecus*. Remains of these near-men have been found in China, Java and Africa.

Above: An artist's view, based on skulls that have been found, of what *Australopithecus* (which means 'southern ape') looked like. These near-men lived in Africa.

It has taken Man only a few thousand years to evolve from living in caves like this . . .

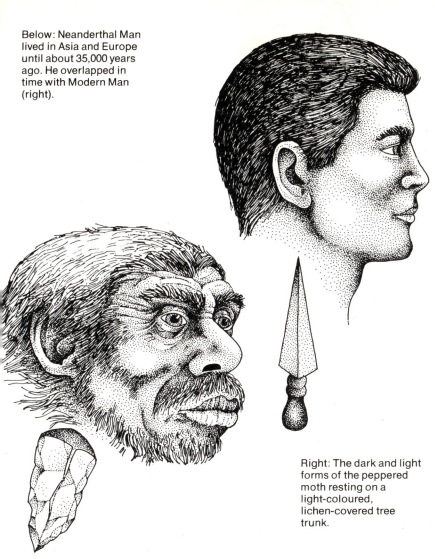

Below: Neanderthal Man lived in Asia and Europe until about 35,000 years ago. He overlapped in time with Modern Man (right).

Right: The dark and light forms of the peppered moth resting on a light-coloured, lichen-covered tree trunk.

...to his present very comfortable life in a house. Yet he still likes sitting by a fire.

these light moths escape being eaten and live to produce young. Occasionally, through a process known as mutation, peppered moths are hatched which are dark grey. They are easily seen on the light tree bark and eaten. However, in sooty towns the bark of trees is not light, but dark. There the light-coloured moths are easy to see, and they do not survive. In dirty surroundings it is the dark grey moths which blend into the landscape and escape being eaten. So now two forms of peppered moth have developed, each suited to its own surroundings. The process of natural

selection is usually very slow. Mutation, however, is a sudden change of characteristics from one generation to the next, and is not always beneficial to a species.

Modern Man is a species known to scientists as *Homo sapiens.* He is the latest species of a group called Hominids, human animals. Nobody knows when the first Hominids evolved. Some of the oldest remains discovered are the skulls of five adults and two children who lived in Ethiopia about 3,000,000 years ago. They belonged to the genus (small group) of Hominids known as *Homo,* true Men. Also in Ethiopia scientists have found the skeleton of a girl who lived at about the same time as these early people. She apparently belonged to another group known as *Australopithecus*—near-Men.

Homo and *Australopithecus* were possibly both descended from the oldest Hominid yet known, called *Kenyapithecus* because part of its skull was found in Kenya.

Nobody can be sure how long *Homo sapiens* has been around, but it seems likely that the period is between 200,000 and 300,000 years. Before *Homo sapiens* there came two earlier types of Man, known as *Homo erectus* (upright Man) and *Homo habilis* (Man with ability). They had smaller brains than *Homo sapiens* and differed slightly in other ways. A later type, *Homo neanderthalensis* (so-called because remains were first found at Neanderthal in Germany), lived in Europe at the same time as modern Man, but died out about 35,000 years ago.

55

Human Body Inquiry Desk

How do people keep their balance?
By using their ears! Your ear is concerned not only with hearing but it also helps you to keep your balance. In the inner ear there are five fluid-filled chambers. These consist of three semi-circular canals and two other chambers, the utriculus and sacculus. The three semi-circular canals lie at right angles to each other. When the body changes position the fluid in the canals moves hairs at the ends of the canals. These hairs contain nerve fibres, which carry messages to tell the brain of the change in position. The sacculus and utriculus indicate any change in the body's position relative to gravity. They also contain hairs connected to nerves, which are surrounded with a jelly-like substance containing crystals of calcium carbonate. As the crystals move, nerve impulses are sent to the brain. These impulses, together with visual information from the eyes, let you know whether you are standing on your head or on your feet. See the diagram below.

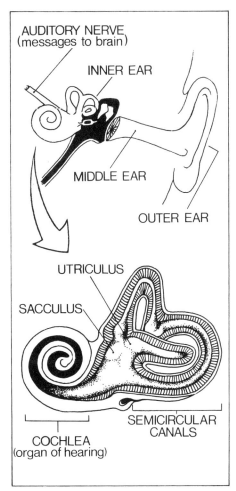

AUDITORY NERVE
(messages to brain)

INNER EAR

MIDDLE EAR

OUTER EAR

UTRICULUS

SACCULUS

COCHLEA
(organ of hearing)

SEMICIRCULAR
CANALS

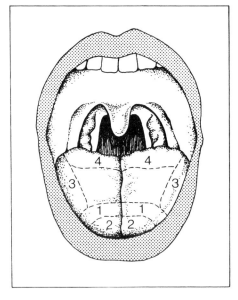

? ? ?

What are taste buds? (above)
They are tiny nerve endings—about 10,000 of them—which are found mainly on the tongue. Some are situated on other surfaces in the mouth. Taste buds respond to four different sensations—bitter, salt, sour and sweet. When the taste buds are activated, which can only happen with liquid or when food has been moistened with saliva, information is sent to the taste centres in the brain.

Different parts of the tongue are sensitive to different tastes. Saltiness is detected all over the tongue, but mainly at the front (1). Sweetness is felt at the tip (2), sourness at the sides (3) and bitterness at the back of the tongue (4). The centre of the tongue has no sense of taste. The senses of smell and taste are closely linked, which is why you lose your sense of taste when you have a bad cold and a blocked nose.

? ? ?

Do women grow old faster than men?
No. Although no one is absolutely certain what causes aging, both men and women show signs of aging from about the age of 30. The aging process follows the same pattern for men and women. The only exception is fertility —the ability to produce children. Women cease to be fertile somewhere between the ages of 45 to 55. Men, however, may stay fertile until well into their 80s.

Both men and women live much longer than people did 100 years ago. Before 1900 men, on average, could be expected to live to between 40 and 45 and women to between 45 and 50. Today, because of improvements in medicine and living standards, average life expectancy for men is between 65 and 70 and for women between 70 and 75. Death rates are higher for men, so in general women tend to live longer than men.

? ? ?

How does the body control its temperature?
Body temperature is controlled by a small structure in the brain known as the hypothalamus. This adjusts body temperature in response to sensations of hot and cold felt by the skin. Normal body temperature is 36.9°C. When the body becomes overheated blushing of the skin occurs. In blushing blood vessels in the skin widen, and blood is carried to the surface. As it is cooled by the air, heat radiates away. At the same time sweat glands in the skin open releasing water, which also takes away heat. When the body gets too cold the blood vessels in the skin contract allowing less blood to reach the surface. This makes the skin look 'blue'. Hairs on the skin may also stand on end and 'goose-pimples' may also appear.

? ? ?

What causes deafness?
There are two main types of deafness. In the first kind sound is actually prevented from reaching the inner ear. This is known as conductive deafness, and it can be caused by a blockage such as wax or fluid in the ear. It can also be caused by a perforated eardrum, an abscess or a bony growth in the middle ear.

Perceptive deafness is more common, and is caused when the inner ear is either damaged or incompletely developed. This may happen because of an illness, such as mumps or meningitis, a head injury, aging or simply by being exposed to very loud noise for too long.

? ? ?

How is food digested? (right)
Digestion is the process in which food

is broken down by chemical agents known as enzymes into carbohydrates, fats and proteins that can be used in the body. It takes place in the alimentary canal—a coiled tube about 10 metres long which runs all the way through the body from the mouth to the anus.

Digestion begins in the mouth where food is ground up by the teeth and mixed with saliva. Saliva helps to break down starches into sugar. The food is then swallowed and passed down into the oesophagus. Muscular contractions force the food into the stomach. Here food is mixed with gastric juices containing hydrochloric acid, protective mucus and pepsin, which breaks down proteins. The food is reduced to an acid pulp known as chyme, which passes into the duodenum, the beginning of the small intestine. Digestive juices from the pancreas and gall bladder complete digestion. The digested food is absorbed into the blood and lymphatic cells. Waste is passed through the large intestine out of the body.

? ? ?

Can two people ever have the same fingerprints? (above)
No. Many millions of fingerprints have been taken, but no two have ever been found the same. Skin consists of two layers of tissue—the outer

layer or epidermis and an inner layer known as the dermis. The dermis contains nerves and blood vessels. Its upper surface is uneven and forms ridges which show through the exactly-fitting epidermis. On the fingers these ridges form fingerprints.

Fingerprints never change because the upper surface of the dermis never changes and each person has a unique set. Because of this fingerprints have been used in crime detection ever since the 1890s.

? ? ?
What is colour blindness?
Colour blindness is the inability to see the difference between certain colours. Three types of cone cells in the retina of the eye are each sensitive to one of the primary colours—red, green and blue-violet. In normal vision these colours are mixed to produce all the colours of the spectrum, but in colour blindness the colours do not appear distinct from each other.

The most common confusion is between green and red. When red light rays strike the eye of a colour-blind person they stimulate both the red-sensitive and green-sensitive cones. The result is a greyish-yellow colour. The same thing happens when green light rays strike the eye. People who are colour blind in this way do, however, learn to tell the difference between red and green because the two colours shine in a different way. There is also a very rare kind of colour blindness in which the eye can only see shades of black and white.

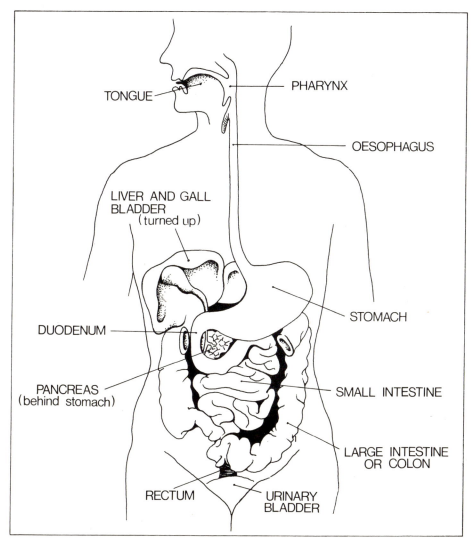

TONGUE

PHARYNX

OESOPHAGUS

LIVER AND GALL BLADDER (turned up)

STOMACH

DUODENUM

SMALL INTESTINE

PANCREAS (behind stomach)

LARGE INTESTINE OR COLON

RECTUM

URINARY BLADDER

The Food Chain

If there were no plants, none of us would be here today. Since animals cannot make their own food, but have to rely on eating other animals or plants, ultimately all our food comes from plants. A plant can trap the Sun's energy and use it to make the complex substances that combine to build the plant's body. Some of the energy may be needed for other purposes, and it can be released from these complex substances when they are broken down into simpler ones again in the process called respiration.

Animals cannot make these complex substances. Herbivores—plant-eating animals—get their complex substances from the plants, then use them either to build up their own bodies, or as energy sources to heat their bodies or move their limbs. Some of these herbivores are eaten by carnivores—meat-eating animals—and so the energy stores and building blocks are passed on.

A feeding sequence of this kind is called a food chain. For example, a leaf is eaten by a

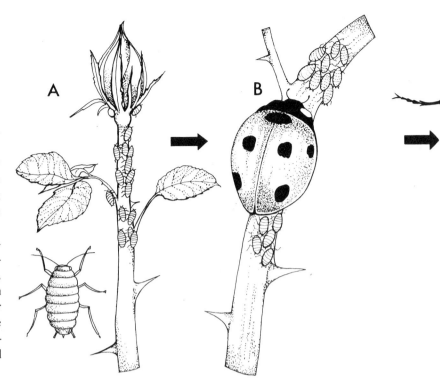

caterpillar, which is eaten by a bird, which is then eaten by a cat, which is eaten by a fox. Or, in the sea, tiny floating plants are eaten by tiny floating animals, which are eaten by small fish, which are eaten by larger fish, which are eaten by even bigger fish.

What happens when plants and animals die? Is all the energy and material locked up in their bodies for ever? The answer is no, because there are many animals, fungi and bacteria which live on the dead remains of other creatures. They break down the corpses and release the nutrients back into the soil. Without these scavengers not only would the Earth become littered with dead bodies, but the soil and the sea would have run out of nutrients long ago.

This, then, is the natural course of events, but when Man interferes by growing crops, the system is upset. When a farmer harvests a field of wheat, for example, the dead plant material is not returned to the soil, and if too many crops are taken from the same patch of soil in quick succession the soil becomes infertile. Early Man moved from patch to patch, leaving the soil to be replenished by natural means over many years, but modern Man needs too much food too quickly. In arid parts of the world the surface soil may blow away between the harvesting of one crop and the planting of the next.

The delicate balance between hunters and hunted, plant growth and animal life that we see in nature has come about by the process of evolution. Every animal and every plant is

Left: Hereford beef cattle—an example of the way in which animal breeders 'improve' animals to get better yields. The Hereford is one of the most popular breeds for beef.

Opposite page: An example of plant breeding—cherry trees thickly covered with blossom, indicating a heavy crop to come, in an orchard in Washington State

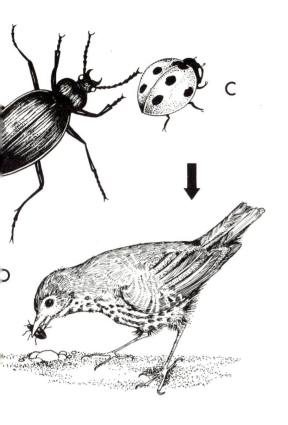

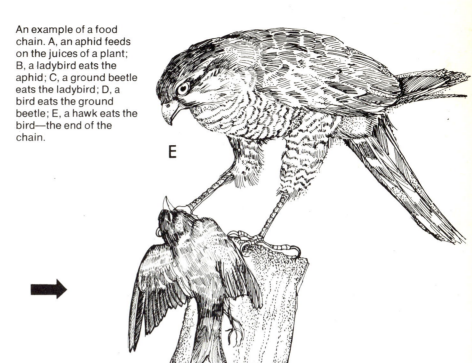

An example of a food chain. A, an aphid feeds on the juices of a plant; B, a ladybird eats the aphid; C, a ground beetle eats the ladybird; D, a bird eats the ground beetle; E, a hawk eats the bird—the end of the chain.

slightly different from all the others just as human beings are all slightly different. Like human beings, too, some are more successful in life than others. Over a very long time, the more successful animals and plants tend to live longer and produce more offspring. In this way, slow changes have taken place in plants and animals, and over millions of years have led to the variety of species you see today.

Farmers make use of this evolutionary principle in breeding domestic animals. By mating together only the best animals in a herd, a farmer can gradually improve his stock, giving for example, cows that yield more milk, or pigs that produce leaner bacon. In the same way plants can also be improved. The plants grown today yield much heavier crops than the wild plants from which they are descended. Although plants have been improving for hundreds of years, the biggest improvements have taken place in the past hundred years. In what has been called the 'Green Revolution' plant breeders have improved some basic crops so much that they have gone a long way towards solving the problem of feeding the world's population.

Japanese growers improved wheat in a series of experiments beginning in 1875. In 1947 the Mexicans took over and by 1968 were producing crops double the size of previous harvests. The new wheat went a long way toward ending the spectre of famine in India and Pakistan in the 1970s. In the same way, plant breeders in the Philippines improved rice in the 1950s and 1960s.

The World of Ideas

Early men's thinking concentrated mainly on the practical day-to-day business of living. Hunters, fishers and food gatherers had to be alert to changes of season and the movements of their prey. They had to react quickly to the destruction of their homes by storm, to the drying up of their water supplies, to the invasion of their territory by other men or by fierce animals, and to famine and plague.

Yet men's minds already brimmed with more abstract ideas long before they became civilised enough to build towns or raise crops from seeds. People looked in wonder and fear at the powerful forces that surrounded and dominated them—thunder, lightning, rain, wind, volcanic eruptions, sun, moon and stars. Even echoes and shadows were awesome things. Men saw all these mysterious forces as living superbeings and worshipped them as gods.

As families grew into tribes they appointed special people such as witch doctors to try to interpret the will of the gods and to protect the tribe from their anger. The witch doctors also became the leaders and custodians of ideas. They combined simple science with superstition and controlled people's behaviour by building up a code of tribal law. The form of the chief god varied from place to place. In regions lashed by strong winds and rain he might be the Storm God; but islanders often created a Sea God. Tribes that had perpetually to fight to keep their territory often worshipped a God of War. Early gods could be seen and touched; they were usually made of forest products or of stone, and later of metal.

The witch doctors took charge of new ideas as these developed. They incorporated them into the tribal religions of which they were the keepers. The earliest-known peoples also believed in an after-life.

Ideas led to the development of increasingly complex styles of visual art, dancing and music. These arts were already highly developed in the early civilisations of Mesopotamia, Egypt, India, Greece and China.

The great leap forward in the communication of ideas came when each civilisation developed a written language. One man's brilliant idea could then be communicated rapidly throughout an empire. Spies and translators could spread the idea into other empires. The idea of the alphabet, for example, beginning in Phoenicia-Palestine, eventually spread throughout most of the world. Only China and Japan kept their non-alphabetic scripts.

Once written down, ideas could be critically examined and rejected or adopted and

A witch doctor near Mbabane in Swaziland. In primitive societies such people were the men of ideas and the leaders of the people's religion.

improved upon. Large numbers of scribes set to work to record and organise knowledge and beliefs, which grew rapidly.

At first the scribes either wrote on damp clay with a stick, scratched on bone, or cut into stone with a chisel. In Egypt the papyrus plant soon provided a more easily-worked material. Elsewhere scribes wrote on parchment, silk, bamboo and leaves. Paper, invented in China about AD 100, later superseded all other writing materials throughout the world. Paper made the scribes' work easier, but every word still had to be written by hand.

The next great improvement in communication came with printing. Block printing (similar to rubber-stamping) was practised in China about 2,200 years ago, but it took another 800 years before block printing reached a high

60

Totem poles like this were carved by American Indian tribesmen to display their family or clan totem (emblem). The totem was considered to be sacred.

standard. Movable type printing was invented in China and Korea in the AD 1300s, but printing became really important when Johannes Gutenberg, a goldsmith from Mainz, set up the forerunner of the modern printing press in Germany about 1440. Printing speeded the communication of ideas and sparked off changes in religion and government.

During the last 150 years the means of communication have multiplied. To books and newspapers have been added the telegraph, telephone, cinema, radio and television. In the mid-1800s news took four months to travel between Europe and Australia. Now, through radio, important events can be known throughout the world within seconds of their happening. The physical barriers to communication have been broken.

61

The Great Religions

The main theme of religion in early civilisations was the worship of powerful forces in nature. Although the Egyptians worshipped more than 2,000 deities, each responsible for some aspect of daily life, their chief god was the Sun, who took the forms of the gods Ra and Osiris. Later, these two gods fused into one. The Sun also became a leading god in Japan, India, Central and South America, Europe and Africa. However, when Pharaoh Akhenaton tried to make the Sun the only god in Egypt (about 1370 BC) his work was quickly undone by the priests. People were not ready to accept the new idea of one god to the exclusion of others.

Perhaps even older and more powerful than the idea of the Sun God was the idea of the Earth Mother. This seemed to explain the great mystery of birth and death. People noted that vegetation 'died' in winter to be 'reborn' in spring. They saw similarities between the fertility that sprang from the soil and the birth of humans and animals. Early men never thought that death was the final end. A dead person was thought either to be reborn into another world or to re-enter this world as another living being.

The Earth Mother represented rebirth and fertility generally. She appeared in almost every religion. In Egypt she was the goddess Isis; in Babylonia, Ishtar; in Greece, Hera.

Egyptian gods with offerings of food, pictured in a mortuary (death) temple dedicated to the great Pharaoh Rameses II.

Stonehenge in southern England was built as a temple, perhaps for Sun worship, about 1800 BC

The 'husband' of the Earth Mother was the Sky Father, a god often identified with the Sun. The Sky Father in Greece was Zeus; in Rome he became Jupiter. The idea of a divine mother and father seemed natural to early peoples because humans lived in families.

As civilisation developed, the various religions arose, with fairly precise rules of conduct for their followers. Most religions had priests and officials, holy writings and special buildings for worship containing images of gods.

Hinduism, the oldest of the great religions, has no known founder. It probably developed as an amalgamation of many local Indian religions. Although Hindus worship many deities — traditionally 330,000,000 — they regard them as only different aspects of one Supreme Being. Hinduism has a vast collection of sacred writings called the *Vedas*.

Buddhism, founded by Siddhartha Gautama (later called the Buddha, 'the Enlightened One') about 2,500 years ago, began as an offshoot of Hinduism. It has no gods, but its followers worship images of the Buddha (often plated with gold) and try to follow the code of behaviour that he advised. Like Hindus, Buddhists believe in reincarnation, return of the soul to Earth in a different body.

Just about the time that Buddhism came into being, four other religions either began or underwent drastic change: Jainism, Confucianism, Taoism and Zoroastrianism.

Jainism, which had similarities with Hinduism and Buddhism, was taught in India by a preacher named Mahavira. In China the philosopher Confucius laid down rigid rules for polite behaviour that later formed the basis of the religion of Confucianism. His teachings were directly opposed by the ideas of Taoism,

Worshippers at a Muslim mosque—the Mosque of Omar Ali Saifuddin in Brunei, Borneo. Note the shoes on the steps— Muslims always remove their shoes before they enter a mosque.

A statue of the Buddha more than 15 metres tall stands in the Temple of a Thousand Lights in Singapore.

said to have been founded by a wise man named Lao Tzu. Taoism taught that men should work in harmony with nature.

The Chinese developed a complex religion that combined Confucianism with Taoism and Buddhism. Japan also adopted Buddhism, and combined it with the ancient religion of Shinto—a form of spirit worship. Chinese and Japanese religions placed great emphasis on honouring ancestors. In Persia (modern Iran) the prophet Zoroaster founded Zoroastrianism, in which fire, earth and water are sacred.

In Palestine the Jews began to develop their religion, Judaism, even before the time of Confucius and the Buddha. The Hebrew Bible began to be written more than 3,000 years ago and took about 1,000 years to complete. More than any other people, the Jews believed in one all-powerful God, whom they could speak with but not see or touch. They also believed that God would send to them, as his 'chosen people', a Messiah (Saviour).

Nearly 2,000 years ago a humble Jewish carpenter, Jesus of Nazareth, began teaching new religious ideas in Palestine. Some said he was the promised Messiah, but Jewish leaders rejected his teachings and persuaded the Roman governor of Palestine to crucify him. The teachings of Jesus developed into the religion of Christianity named after Christ, the Greek word for Messiah, and the title people gave to Jesus.

Christians adopted the Jewish Bible (which they called the Old Testament) and added the New Testament to it. Unlike Judaists, Christians believe in eternal life after death. They believe that men's sins were 'redeemed' by the shedding of Jesus's blood. Like Buddhism, Christianity split into sects.

Islam drew inspiration from both Judaism and Christianity. It dates from the year AD 622, when Muhammad, an Arab merchant who claimed to have revelations from God, fled from Mecca to Medina in Arabia. Islam means 'submission' to the one God (Allah). Islam spread quickly throughout the Arab lands. Later it spread into black Africa and eastwards as far as Indonesia. Strict rules govern the conduct of Muslims, as the followers of Islam are known.

Sikhism was founded by Nanak, a Hindu *guru* (teacher), a little over 500 years ago, just about the time that western Christianity split into the Roman Catholic and Protestant Churches. Founded in the Punjab, India, it combined aspects of Hinduism and Islam.

The major international religions today are Christianity, Islam and Buddhism. Hinduism, Taoism, Shinto, Judaism and Sikhism remain national religions. A few Jains and Zoroastrians still live in India. Confucianism has died out, but has left its mark on Chinese culture.

The Great Philosophers

Religions are ideas about the unseen power of God. People can accept these ideas as a matter of belief without thinking very strongly about them. However, many people devote their lives to thinking very deeply indeed about the problems of life. Such people are the philosophers. Some philosophers, such as Confucius, have had religions founded on their teachings, but others have stood outside the main religions in their thinking.

Greek philosophy began to develop in the 500s BC, about the same time as the great upheaval of religious ideas in India, China and Persia which is described on pages 62-63. One of the greatest Greek philosophers was Socrates, who taught by question and answer. He shook people's beliefs in formerly accepted things, especially Greek religion, and was forced to commit suicide as a punishment. His death failed to stem the flow of new ideas in Greece. Greek scholars turned their brilliant minds to devising new systems of government, to science and mathematics and to new styles of art and architecture. The Greeks argued about such questions as 'What is truth?' The questions they raised are still debated in our time.

Although the Greek ideas were brilliant, they contained many errors of fact. Yet Greek thinking still dominated the minds of people in medieval Europe nearly 2,000 years later. Then about 600 years ago came the European Renaissance (Rebirth of Learning). Old ideas began slowly to crumble away and make way

The Greek philosophers Socrates and Plato, as pictured in a manuscript of the Middle Ages. Plato was Socrates's pupil.

Reading the Declaration of Independence in the city of Philadelphia on July 8, 1776, from an old engraving. This was the first occasion on which the public had a chance to hear the Declaration.

for new. Science began to oust superstition. In France René Descartes (1596-1650), the 'father of modern philosophy', devised a new method of thinking. He said: 'I think, therefore I am'—in other words, 'if I am able to think, I must exist'. Beyond that he accepted nothing unless and until he could see its truth 'clearly and distinctly'. Everything had to be doubted until proved. Big problems, thought Descartes, were really multi-problems. They should be divided into as many separate problems as possible and each small problem could be solved relatively easily.

Thomas Hobbes (1588-1679), who lived through the English civil war of 1642-46, favoured strong government. According to him, the lives of men had been 'nasty, brutish and short' until they agreed to form themselves together under a social contract and surrender power to a king. Men could not rebel against their king, said Hobbes, because that would be rebelling against themselves. This did not convince the English Parliament, which cut off the king's head. Neither did Hobbes please the royalists, who claimed that a king's power came from God.

John Locke (1632-1704) did not agree that the lives of primitive men had been 'nasty, brutish and short'. He thought that God had created men with a divine spark to become naturally reasonable beings. Men brought 'natural rights' into society when they agreed to join it, said Locke. Therefore, the king held power only as a trustee of those rights. Locke supported the English 'Glorious Revolution' of 1689 which deposed King James II, because he said that James had offended against

Above: The great French philosopher René Descartes. Right: Karl Marx and his friend and colleague Friedrich Engels discussing politics with socialists in Paris in 1844. Four years later the two men produced the *Communist Manifesto*, the basis of modern Communism.

Vladimir Lenin addressing men of the Red Army in Red Square, Moscow, in May 1919.

'natural law' and so lost his right to be king. Locke's ideas pleased the English Parliament, which reduced the power of monarchs.

The men who drew up the American Declaration of Independence in 1776 put Locke's ideas into more down-to-earth terms. They wrote: 'We hold these truths to be self-evident, that all men are created equal, that they are endowed by their Creator with certain unalienable Rights, that among these are Life, Liberty, and the pursuit of Happiness. . . To secure these rights, Governments are instituted among Men, deriving their just powers from the consent of the Governed.'

In 1848, when many European peoples rebelled against their kings, Karl Marx, a German writer, issued the *Communist Manifesto*. In it he declared: 'Let the ruling classes tremble . . . the proletarians [workers] have nothing to lose but their chains.' Later, Marx claimed in his book *Das Kapital* (Capital) that a study of history showed that the ruling class had always exploited the proletariat (working class). The power of the ruling class to exploit, said Marx, came from its possession of land, mines, factories, houses, shops and other forms of 'capital'. So long as 'capitalists' controlled the 'means of production', said Marx, class war was inevitable.

At that time the old ruling class of landlords was giving way to the new middle class of industrialists—the bourgeoisie. It was this class especially, said Marx, that the proletariat had to defeat. Marx advocated Communism—a society in which all the means of production would be owned by the state.

In 1917 Lenin established the world's first Communist or 'Marxist' state in Russia. Sixty years later nearly two-fifths of the world's peoples lived in Marxist states. Politics has now replaced religion as the main arena for the clash of ideas.

Myths and Legends

Alongside religion people have developed myths and legends. Myths are complex stories told to explain vital matters such as the beginnings of the world, the origins of a people or country, death or the activities of gods and monsters. Myths passed by word of mouth from generation to generation. They became embellished in retelling, and often became incorporated into religion. Legends are stories that usually develop around real or imaginary people. There is at least a grain of truth in most myths and legends, and no clear borderline exists between them. Myths and legends form an important part of the culture of a society and help to hold it together.

The ancient Greeks had a rich mythology built around the activities of their gods and goddesses. Early myths tell how the first Earth Mother (Gaea) and the first Sky Father

Opposite: A goddess of motherhood, worshipped by the Totonac tribe of Aztec Indians who lived in Mexico before the Europeans conquered the country. The people of that time had a great many gods and goddesses.

Zeus, the chief of the gods in Greek mythology, kidnapping Ganymede, son of a Trojan prince. The gods wanted Ganymede to be their cup-bearer and wait on them at meals.

(Uranus) became the parents of 12 giants called Titans, and several monsters including one-eyed giants called Cyclops. One of the Titans, Cronus, attacked his own father and replaced him as chief god. Then, frightened that he might suffer the fate of his father, Cronus swallowed each of his own children as soon as they were born.

However, one of his children, Zeus, escaped because some of the goddesses put a large stone in front of Cronus and tricked him into swallowing it in the belief that it was Zeus. As had been predicted by an oracle (a divine statement made at a holy place), Zeus eventually toppled Cronus from power. He forced Cronus to cough up the stone, and along with it came all Zeus's swallowed brothers and sisters. Cronus was banished and Zeus claimed to be the new chief of the gods. Great battles then took place between the Titans and the other giants who supported Cronus and the old gods, and the new gods led by Zeus. Zeus won.

These early Greek myths resemble those of Hinduism, and the two sets of myths probably had a common origin. The three generations of gods are thought to symbolise different stages of Greek civilisation.

Under Zeus lived a vast collection of deities with human-like behaviour and emotions. When Greece came under Roman rule the Romans adopted Greek deities under different names. For example, they called Zeus Jupiter, and Cronus was Saturn. In the next paragraph the Roman names are given after the Greek names each time a new deity is mentioned.

Zeus (Jupiter) gave the underworld, or hell, to his brother Hades (Pluto). The sea became the kingdom of another brother, Poseidon (Neptune). Zeus kept the sky for himself. He lived with his rebellious wife Hera (Juno) at the top of Mount Olympus, in northern Greece. Zeus's sisters lived there too—Hestia (Vesta), the kind goddess of the hearth-fire, and Demeter (Ceres), goddess of agriculture. Eight other deities lived with them: Poseidon; Aphrodite (Venus), goddess of beauty; Athene (Minerva), goddess of wisdom, who protected things; Apollo (Apollo), the handsome god of song and music, who also destroyed things; Artemis (Diana), goddess of hunting; Hera's son Hephaestus (Vulcan), the blacksmith; Ares (Mars), god of war; and Hermes (Mercury), a winged practical joker who flew everywhere as messenger for Zeus.

The Vikings, and Germanic peoples such as the Anglo-Saxons, shared a mythology which was part of their religion. Odin (or Woden) was god of heaven and earth, agriculture and war. Thor, son of Odin and his wife Frigga, wielded a mighty hammer from which flashed lightning. Thunder was believed to be caused

by Thor's chariot rolling across the sky. Woden's day is still celebrated as Wednesday in Britain, and Thor's day as Thursday. Latin countries such as Italy still name some days after the Greco-Roman deities, as we do for Saturday (Saturn's day).

King Arthur, the hero of many English legends, was said to have lived about AD 500, when the Romanised Britons had been driven into Wales and Cornwall by Anglo-Saxon invaders. Some legends place his court at Camelot. Arthur's knights sat with him at the Round Table, but his chief assistant was Merlin, a wise magician. Arthur was probably based on a real British resistance leader, though not a king. He served as a symbol of the opposition of the Britons to Anglo-Saxon rule. Medieval knights respected him as a man

One of the most famous Greek legends tells of the siege of Troy, a city which stood in what is now western Turkey. A Greek army spent 10 years trying to capture it but failed. Then the Greeks pretended to go away, leaving behind a huge wooden horse. The people of Troy hauled the horse inside the city walls to see what it was—and at dead of night Greek soldiers hidden inside crept out and opened the gates. The rest of the Greek army, which had been hiding, came in and captured the city. The story was told by the poet Homer, about 800 BC; this picture was made by the Venetian artist Giambattista Tiepolo in the AD 1700s.

devoted to chivalry, and the protection of the poor and weak.

Robin Hood was the hero of poor people in medieval England because he and his 'merry men' robbed the rich to feed the poor. Belief in Robin Hood gave poor people an outlet for their hatred of kings, officials, bishops and other rulers who oppressed them.

To the medieval Swiss, William Tell was a legendary hero who dared to defy the Austrian Habsburg rulers of Switzerland. When Tell refused to bow to an Austrian hat set on a pole, he escaped punishment only by agreeing to shoot an apple off the head of his own son. Although Tell did this without injuring his son, he was seized on another charge. Tell lived to lead a Swiss revolt against the Austrians.

In Japanese mythology the deities who

The ancient Persians had many legends about Rustem, a great hero who was known as the 'Shield of Persia'. This beautiful miniature painting shows Rustem, sleeping, being saved from a lion by his great yellow horse Rakush.

Below: William Tell, crossbow on his shoulder and one arm around his little son, as shown in a monument to his memory at Altdorf, in central Switzerland. This town was, by tradition, where Tell shot the apple off his son's head.

heaven and took on the form of the god Vishnu. This myth, told in the Indian epic *Ramayana*, is extremely important in Hindu culture.

A myth from Madagascar gives an explanation for life and death. God's daughter Earth amused herself by making little men out of clay. When God saw the clay figures he breathed life into them and named them Velo (Living). The Velo men increased in numbers and worked on the land to produce good crops and make Earth prosperous. But God grew jealous and demanded half the Velo men. When Earth refused to surrender them God took away the breath of life. Since then, men grow old and die.

Africa has several myths about twins. In Mozambique they were called 'children of the sky'. Their birth was seen as a disaster to be put right only by purification rituals, and they and their mother were banished outside the village. However, when storms came twins were expected to plead to their Sky Father for the safety of the village.

The American poet Henry Wadsworth

controlled heaven sent down the god Izanagi and the goddess Izanami to create land on the great sea of the Earth. Taking a jewelled spear, they stirred up the sea, and drops of water that fell off the spear solidified into the first island of Japan. Later, Izanami gave birth to the other islands of Japan.

Another Japanese myth tells how Amaterasu, the Sun goddess, fled into a cave after a fright, bringing darkness upon the land. The other deities eventually tempted her out by setting up a mirror near the cave entrance. When Amaterasu peered out to see her reflection, the God of Force grabbed her and made her stay out of the cave to light the world. The myth probably symbolises the return of spring after winter. Traditionally Amaterasu is the ancestor of the present emperor of Japan, and a mirror still forms part of the Japanese royal regalia.

An Indian myth tells how Rama, an Indian prince banished through trickery, was exiled for 14 eventful years in the forest. Sita his wife and Lakshmana his loyal half-brother shared his exile. By another trick the ten-headed demon-king Ravana kidnapped Sita. With the help of Hanuman, the monkey king, Rama eventually captured Sri Lanka, Ravana's island kingdom, and returned home to inherit his father's kingdom in triumph; but Rama doubted Sita's faithfulness to him during her captivity. He banished her, and in despair she called upon her Earth Mother to swallow her up. In remorse Rama ascended to

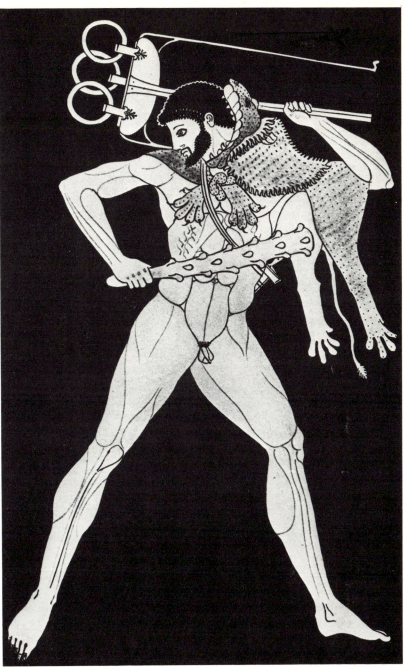

Longfellow collected many stories about various gods and heroes of the Chippewa Indians and incorporated them into his long poem *Hiawatha*. In fact Hiawatha was a real person—a peace-loving chief of the Mohawk tribe who lived about 400 years ago. Longfellow's poem turned him into a legendary figure.

Movements of the Sun, Moon and stars were important to the Eskimos because, being hunters and fishers, they needed to be able to know the time of the year. The Eskimos thought of the Sun and Moon as sister and brother running a race. At first the Moon keeps close to his sister, the Sun, but she gradually overtakes him.

The Eskimos thought there were many spirits that helped them, but they had no real gods. However, they believed in a giant Earth Mother called the Old Woman, who lived under the sea. As a child she became so hungry that she began to eat her father and mother, who threw her into the sea in terror. The Old Woman became the mother of all things living in the sea and was also the cause of storms at sea.

A legendary hero of more recent times was Paul Bunyan, who symbolised the lumber cutters and frontier men of the American Wild West. Paul Bunyan's supposed exploits combined the boasts, jokes and tall stories associated with real-life folk heroes such as Davy Crockett. Everything about Paul Bunyan was larger than life, including his enormous blue ox, Babe, which needed the waters of the Great Lakes to quench its thirst.

Above left: The Lady of the Lake appears to King Arthur and tells him of the magic sword Excalibur. This picture was drawn by the English artist Aubrey Beardsley (1872-1898). Above right: A picture of the Greek hero Heracles (Hercules), painted on a vase. Heracles performed twelve apparently impossible labours.

Although there never was a real Paul Bunyan, the legends surrounding him served to illustrate the bigness of the United States.

A well-known mythological character among Polynesian peoples was Maui, known as Maui of the Thousand Tricks, Maui Super Superman, or Maui the Wise. Maui was the ugly baby of a human mother and a divine father. Although the gods brought him up, Maui preferred the company of humans and took great risks in using his supernatural powers to help them in their early days. He raised the sky, stole fire from the gods, slowed down the Sun, subdued the winds and set the stars in their places. However, his achievements often brought disaster and his clowning lost him the support of gods and men. Before he finally met death he made one last great effort to gain immortality for men, but failed.

Sharing Ideas

An idea is generally of little value unless it can be shared with other people. The story of the modern world is basically the story of communications—the art of sharing ideas. As you can see on pages 60-61, the first great steps in communications were the development of language, writing, and printing—the multiplication of the written word. Language has existed for many thousands of years, writing came into being about 5,000 years ago, and printing was developed about 1440.

For a long time there were two basic prob-

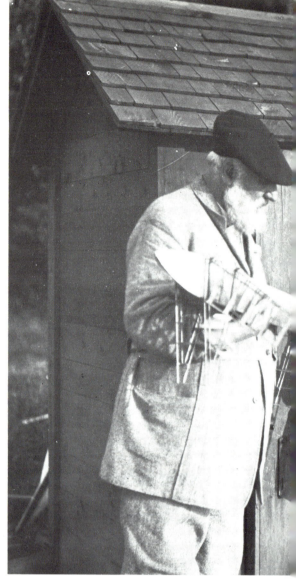

Below: A telegraph machine invented by David Hughes (who later invented the microphone) in 1855. The machine was made before the typewriter keyboard was invented, so Hughes adapted the only keyboard then in use—that of the piano. His machine remained in use until the 1930s. Bottom: A telegraphic instrument of the type Charles Wheatstone invented in 1840. To send a message you cranked the handle to generate electric power.

lems of communication: distance and time. These are probably best demonstrated in the War of 1812 between Britain and the United States. The war was officially ended by a treaty signed at Ghent in Belgium on December 24, 1814—but because the news travelled so slowly one of the bloodiest battles of the war was actually fought at New Orleans 15 days later. People could exchange ideas and news only as fast as a ship could sail or a man or a horse could run.

Over short distances people began using semaphore, signalling with flags or with movable arms on a tall post. One of the most successful semaphore systems was invented by a Frenchman, Claude Chappe, in 1791. The discovery of electricity led people to experiment with means of using it for transmitting signals.

Real progress with telegraphy came in the 1800s when better sources of electric power were available. The first successful electric telegraph was invented by two Englishmen, William Cooke and Charles Wheatstone, in

The inventor of the telephone, Alexander Graham Bell (left), discusses with his laboratory superintendent, Hector P. McNeil, some models of hydrofoil boats designed in 1912.

Modern telephones have press buttons for quick calling of numbers.

did it in 1895. By 1901 he was signalling across the Atlantic Ocean.

Television was the work of many scientists, but a Scot, John Logie Baird, was the first person to achieve a practical demonstration of TV transmission, which he did in 1926. Television's progress was slowed down by World War II (1939-1945), but developments came thick and fast in the 1950s. A snag with TV is that its waves travel in straight lines and do not follow the curve of the Earth, but the development of satellites in the 1960s and 1970s means that TV signals can now be 'bounced' off a satellite, enabling broadcasts to be seen simultaneously all over the world.

1837, and it was followed a year or two later by the even more successful telegraph of Samuel Morse. Morse, an American artist, also invented the code of dots and dashes which quickly came into general use for the electric telegraph.

Pictures were transmitted by wire as early as 1863 in France, but the modern system of sending photographs by wire was not developed until 1907. It was perfected in the 1930s.

The telephone was invented by a Scots teacher of deaf-mutes, Alexander Graham Bell, working in Boston, Massachusetts, in 1876. This invention and that of the telegraph meant that news could be sent over long distances within a few seconds of it happening. However, both the telegraph and the telephone depended on a wire linking two distant places. From the 1870s onward scientists worried away at the problem of sending messages through space, using electromagnetic waves (see pages 174-175). The first person to send radio signals successfully was an Italian scientist, Guglielmo Marconi, and he

BOOKS OF WORDS

Dictionaries have provided a very important aid to communication. They define the meanings of words, so that people can be sure exactly what an author is saying. They also standardise spellings, thus making written language easier to understand for everyone.

The first true dictionary of the English language was prepared by Nathan Bailey in 1721. It was followed in 1755 by Samuel Johnson's *A Dictionary of the English Language*. America's great dictionary pioneer was Noah Webster, who produced his first large dictionary in 1828. Webster was responsible for several differences between British and American spellings, such as theatre/theater, colour/color, sceptic/skeptic.

Producing a Book

The unknown author of *Ecclesiastes* in the Bible, writing about 250 BC, said: 'Of making many books there is no end'. His words are even more true today, when publishers produce thousands of books every year. There are many kinds of books, but they fall into two main groups: fiction (novels and stories) and non-fiction—all the rest.

The idea for writing a novel begins with its author. When he has completed his manuscript he then has to try to interest a publisher in it. If he is a new writer, this may be difficult: some authors try 30 or 40 publishers before they find one willing to accept their work. This is because it costs a great deal of money to print and publish a book, and if the publisher is not sure how well it will sell he may be reluctant to invest that money. Once established, an author has much less trouble, but only a few novelists make enough money from their books to live on.

With non-fiction books the situation is very different. Some, such as a novel, are written first and then offered to a publisher, but most are the result of a definite request from a publisher to an author. Often the author approaches the publisher with an idea, but he does no writing until the publisher has agreed to take the book and give him an advance—that is, a payment on account of the sales that are expected.

Most non-fiction books are written to fill a need. A good example is a school textbook. Books about crafts, places people may want to visit, plants, animals, and other special subjects are some other examples.

There are two main ways of printing a book. The traditional method, used ever since Johannes Gutenberg invented it in the 1400s, is known as letterpress. In this process the letters are made in a mirror-image, back to front, and their surfaces are inked. When paper is pressed on the type, an impression is produced. Type used to be composed (set up) by hand, using individual letters. Today machines set the type. They are operated from keyboards, like that of a giant typewriter.

Letterpress is usually printed on soft paper which slightly absorbs the ink. To print pictures a better quality paper is needed, and this is often a section on its own. Sometimes separate leaves, called plates, are pasted into a book by one edge.

The other main method used for printing books is called lithography, litho for short. Lithography was invented by a Bavarian actor, Aloys Senefelder, in 1796. Senefelder drew on a certain kind of porous stone with a greasy crayon, and then wet the stone with

water. The stone did not absorb the water in those areas where the crayon lines were. He next passed a roller full of oily ink over the stone. The ink stuck to the crayon but not to the wet part, and when Senefelder laid a sheet of paper on the inked stone he printed off a perfect impression of his drawing. Modern lithography uses thin metal plates instead of stone, and the image is put on to the plates by photography, but the basic process is the same. Lithography is a very good method for printing pictures, which is why we chose it for *Purnell's Pictorial Encyclopedia*. Text in litho printing can be reproduced from an impression taken from metal type, but it is more often 'set' photographically, as this book is.

Books are printed on large sheets of paper holding several pages at once. The pages are multiples of eight, usually 16 or 32 to a sheet. The sheet is folded up so that the pages come in the right order, and is then called a gather or a signature. In a large book the signatures are sewn together. The spine (the back of the book) is reinforced with a coating of glue, and the book is finally glued into a hard case.

Today small books, and almost all paperback books, are not sewn. The pages are trimmed so that they form separate sheets, and are then glued along one edge, just like a tear-off writing pad. This is a much quicker and cheaper method of binding, but it does not last so long.

A few expensive books are bound in leather, but most books other than paperbacks have stiff bindings covered with fine, hard-wearing cloth. To keep these covers clean on the bookseller's shelves, the publishers wrap a paper dust jacket around the book. At first dust jackets were very plain, but today they are bright and colourful, and are designed to make the book catch your eye in the shop.

Above: Charles Dickens was one of the greatest writers of novels. In *Dickens's Dream*, the English artist Robert William Buss shows the author in his study with many of his characters.

Opposite, top left: At work with a modern photo-setter, which produces film of the type-matter of a book. Top right: Setting metal type by hand, as it has been done ever since Gutenberg's day. The tray holding the stock of letters is called a case, and the small one in the compositor's hand is called a composing stick.

Opposite, centre left: A printer locks up metal type in a heavy steel frame called a chase. Bottom left: Using a knife and paste to make up a page from photoset material.

Opposite, far right: A letterpress printing press, using ordinary type. Inset: A high-speed press for printing from litho plates.

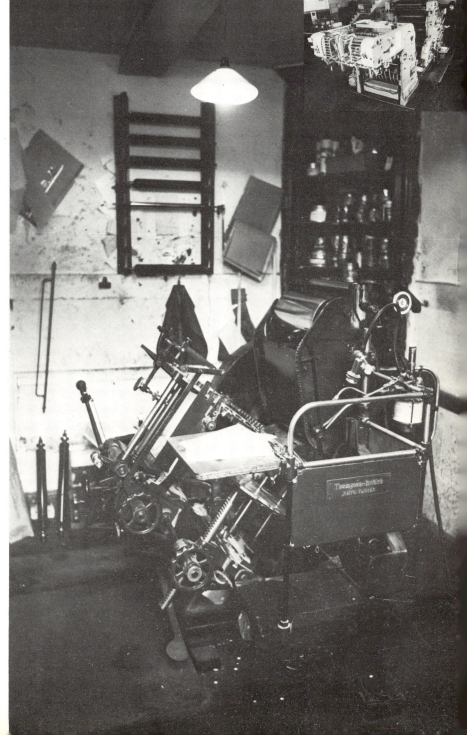

Ideas Inquiry Desk

What is astrology?

Astrology is the study of the Sun, Moon and planets in the belief that they influence and determine the affairs of people. Astrologers take the Earth as a central point. The planetary belt around the Earth, including the Moon, is divided into 12 signs, each of which is 'ruled' by a particular planet (shown right).

In order to describe a person's characteristics and their determining influences, an astrologer draws up a horoscope or birth chart. This shows the exact position of the planets at the time of that person's birth. The pattern of planets that emerges supposedly determines the person's personality or future. There is no scientific reason for holding such beliefs.

Astrology is an ancient practice. It was probably first used by the Chaldeans and Assyrians, and it then spread throughout Asia and Europe. With the rise of Christianity it declined, but became important again in Europe during the 1400s and 1500s. Today it is only seriously practised in the East although it is quite popular with some people in Western countries.

? ? ?

Which is the first book in the Bible, and how many books are there?

Genesis is the first book. It tells the story of Abraham, the father or founder of the Jewish nation. The Bible consists of two parts—the Old Testament and the New Testament. The Old Testament is the oldest part of the Bible and was written mainly in Hebrew. It consists of 39 books which contain the story of the Jews and most of the Jewish teachings. The first five books contain the Jewish *Torah*, or law. These are followed by the Prophets, and the remaining books consist of the Writings—poems, sayings and stories.

The New Testament contains 27 books, written in Greek. The first four are the Gospels which describe the life of Jesus Christ. These are followed by the Acts of the Apostles and the Epistles, or letters, most of which were written by St. Paul. The last book in the Bible is Revelations, a book of prophecies.

In addition to the Old and New Testaments there is a collection of 14 books called the Apocrypha, which means 'hidden writings'. They are not always accepted as divinely inspired, which the other books of the Bible are thought to be.

? ? ?

Who founded democracy?

The word democracy comes from the Greek word 'demos' meaning people, and democracy means power of the people or a system of government in which the people share in making vital decisions. Democracy originated in ancient Athens where, in 594 BC, a statesman called Solon drew up a code of laws and set up a council of 400 members. All citizens, except the very poor, had the chance to serve on this council. Solon also introduced trial by jury in the courts.

Democracy disappeared for a while in Athens, but in 506 BC was reintroduced by the statesman Cleisthenes. He set up a new council, the members of which were chosen by the citizens themselves, and everyone had a say in the government of the city. It was from this idea that modern democratic governments evolved.

? ? ?

In ancient Greek legends what was the Trojan Horse?

It was a large, hollow, wooden horse, and it forms part of the story of the Trojan War. The war began when Paris, son of the King of Troy, captured Helen, wife of King Menelaus of Sparta. Menelaus and the other Greek kings set off to Troy to bring Helen back. They besieged Troy with no success for 10 years, finally entering by a trick.

The Greeks built a huge wooden horse in which a group of soldiers was hidden. The Greek army then pretended to sail away, leaving the wooden horse behind. The Trojans brought the horse into the city. That night the soldiers let themselves out and opened the city gates to the rest of the Greek army. The Greeks then destroyed Troy by fire. This legend was used by the Greek poet Homer for his famous epic poem, *The Iliad*. There probably was a real Trojan war about 1200 BC.

? ? ?

Who was Hercules?
He was the most popular of the legendary Greek heroes. Hercules is his Roman name: the Greeks called him Heracles. He is always shown as a muscular man wearing a lion's skin and carrying a club. The legends say he was the son of Zeus, father of the Greek gods, and Alcmene, a princess. Hercules was renowned for his great strength and courage. Even as a baby he strangled two serpents.

When Hercules became a man he served Eurystheus, King of Tiryns, for 12 years and performed 12 great labours. These included killing the Nemean lion; slaying the Hydra, a serpent with a large number of heads; cleaning the Augean Stables; and capturing Cerberus, the many-headed dog who guarded Hades. Hercules was involved in plenty of other adventures. When he died he was carried off to Mount Olympus, home of the gods, where he received the gift of immortality.

? ? ?

In Arthurian legend, what was the Holy Grail?
The Holy Grail was a cup or chalice that was supposed to have been used at the Last Supper of Jesus and His Apostles, and then taken to England by Joseph of Arimathea. Many of King Arthur's knights went in quest or search of the Holy Grail but according to legend it could only be revealed to a pure knight. There are many different versions of the quest for the Grail. The picture (right) shows King Arthur and his knights in a tournament, and comes from an old manuscript book.

? ? ?

Who invented the telephone?
The forerunner of the telephone as we know it was invented by a Scottish professor, Alexander Graham Bell (1847-1922), who lived in the United States. He was particularly interested in teaching deaf people to talk. He became a professor at Boston University and with an assistant, Thomas

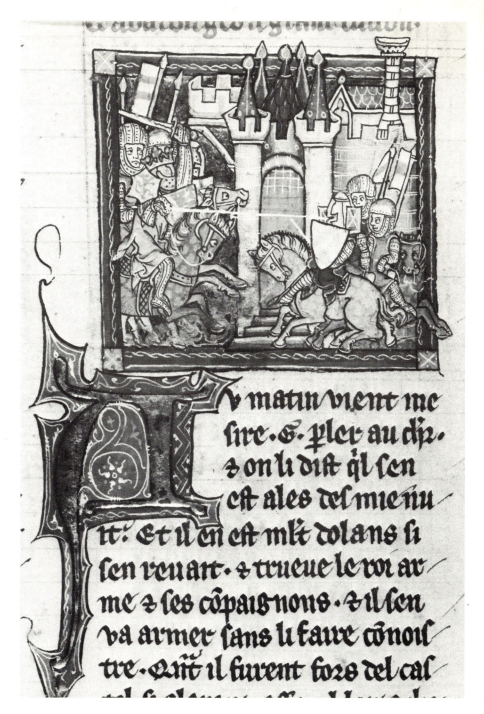

Watson, worked on the idea of transmitting speech by electrical waves. He spent long hours experimenting with a metal transmitting diaphragm suspended near a coil wound round a magnet. In March 1876 he made the first-ever telephone call to his assistant, saying 'Mr. Watson, come here—I want you'.

? ? ?

What is an encyclopedia, and what does the word mean?
An encyclopedia is one of the most useful types of reference works. The term comes from Greek words meaning 'general education', and essentially an encyclopedia contains facts on a selection of the most important subjects. It can consist of either one single volume or several, and it can be arranged alphabetically or by subject, as with *Purnell's Pictorial Encyclopedia*. An encyclopedia is different from a dictionary, which usually contains only the meanings and pronunciations of words.

The earliest existing encyclopedia was written by a Roman, Pliny the Elder. It was called *Natural History* and was completed about AD 77. One of the most famous encyclopedias was the French *Encyclopédie*, which was completed in 1772.

The Story of a Newspaper

Before the development of radio broadcasting in the 1920s and television in the 1930s and later, newspapers were the main way in which people found out about what was going on in the world around them. The first real newspaper was founded in Germany in 1615, and over the next 250 years many papers were established. Most of them were small, and were bought by only a few people. The cheap, popular newspaper with a big circulation has developed since the 1890s, when for the first time most people in western Europe and North America were able to read.

A newspaper provides four main services to its readers. First of all it contains news, some of it reported in great detail, some just mentioned. Second, it gives the readers background information about people and things in the news, so that they can understand what is happening. Third, it provides features, which include all kinds of material not related to the news, such as special articles, crossword

Right: In the features department of a British national newspaper a layout for the next day's paper is discussed. Below: The 'Big Room', with reporters in the foreground, and the sub-editors in the background.

puzzles, strip cartoons, and services such as TV programmes and weather forecasts. These features are often in a special section of the paper; other sections may be devoted to sport or financial news.

The fourth service provided by a newspaper is to carry advertising. Although there are many other ways of advertising, for instance on radio, television and posters, many kinds of goods and services can only be adequately advertised in a newspaper. They include jobs vacant and wanted, small articles for sale, holiday services, and financial matters.

However, advertising plays another very important part in newspapers. Almost all newspapers cost a great deal more to produce than they earn when the copies are sold. The difference is made up by the money the advertisers pay: without that money each newspaper would cost four or five times as much as it does now.

Much of the cost is the result of the speed with which newspapers are produced. To collect news, prepare it for publication and print it all in a few hours means using very big staffs. News comes to a paper from many sources. First are the paper's own reporters, who may be able to gather the news by making telephone calls, but often have to travel to the scene where the news is happening. This may mean a journey of a couple of kilometres to a fire the other side of town, or a flight to a war-torn country the other side of the world. The main source of general news is the news agencies, which have a great many local reporters in all parts of the world. The biggest

Right: A section of one of the battery of high-speed presses which prints a newspaper. The printing is done by curved plates clipped on to a cylinder.

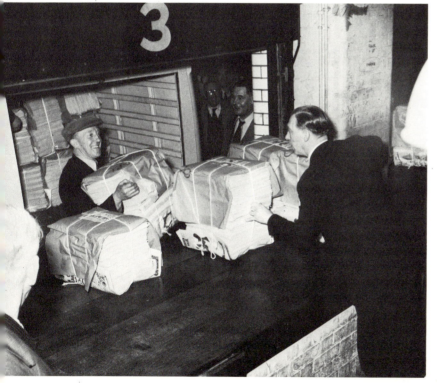

such as the advertisements, features, some of the background information and early news stories, are already in type. In a very few hours the paper is completed and assembled, and plates are made for the printing presses, which can roll off thousands of copies an hour. Even then the work of the editorial staff is not finished, because late news is coming in all the time. Every so often the presses are stopped and fresh plates with the latest news are put on them.

Pictures for a newspaper come from similar sources to the news, and are chosen by a picture editor. A newspaper usually has a large stock of pictures, particularly portraits, so that if some famous person is suddenly involved in the news his picture can be used.

The final part of newspaper production is distribution. Copies of the paper are tied up in bundles and taken by vans to newsagents nearby, or to waiting trains and planes to go to people further afield. In some countries, such as the United States, many copies are posted direct to subscribers.

agencies are Reuter's, United Press International and Associated Press.

When the news arrives in the newspaper office the editor and his production team—called sub-editors in Britain and copy readers in North America—decide how much space each story should receive, write the headlines, if necessary rewrite the stories (often from several accounts) and send the material for typesetting.

The less urgent parts of the newspaper,

Above: The mechanical work is done and it's back to manual labour for the last stage—loading bundles of newspapers into the waiting vans which (right) speed to distribute the papers to the street sellers and to the trains which carry the bundles to more distant places.

Radio and Television

In just the same way as a lot of people and a great deal of hard work are needed to produce a newspaper, so a few minutes of material broadcast on radio or television can mean many hours of work beforehand. Both aspects of broadcasting are similar, but television is much more complicated than radio and involves far more people.

Most radio programmes are made in a studio. This is a room soundproofed so that no unwanted noises can get in to spoil a broadcast. The studio contains one or more microphones, depending on how many people are involved in the broadcast. Next to the studio is a control room, where the producer sits. He can see into the studio through a window. The studio control room also contains machines for playing tapes and discs, and a control panel which enables the producer or one of his assistants to control how loud or soft the sound is from each microphone, and to feed in sounds from tapes or records.

Effects—sounds such as car engines, birds singing or trains going by—are largely on tape or disc, but some, such as the sound of doors opening and closing, the clink of teacups, or footsteps on gravel paths, are made in the studio.

Most broadcasts begin with a script, which sets out what people are going to say. An informal programme such as a discussion has very little scripted matter, usually just the opening and closing remarks. Nearly all broadcasts have one or more rehearsals, so that people know exactly what they are going to do. Even quiz programmes are rehearsed, using different questions from those heard in the actual programme, so that the contestants get used to being in a studio.

Timing is all-important, and one purpose of

A radio studio control room set up for a quadraphonic broadcast, using four channels for the sound output.

In a radio studio for an episode of the BBC drama series *Vivat Rex*. The actors are (left to right), Jeffrey Segal, Sir Michael Redgrave, Nigel Stock, and Clifford Rose, with the director, Gerry Jones.

rehearsals is to make sure that the programme will not be too short or too long. A programme of gramophone records is built up with the exact time of each item carefully noted so that the script for the announcements between each record can be written to fill the time that is left. This timing is not so important with the informal disc-jockey programmes of 'pop' music, in which the DJ makes up most of what he says as he goes along.

Television programmes also take place in a studio, a much bigger one to allow for several cameras. Each camera has its own cameraman, who is able to hear instructions from the director of the programme through headphones. Some cameras also have assistant operators who wheel the cameras about, or raise them up for high-level shots. For a discussion or talks programme the microphones may be 'in shot'—that is, you can see them on the screen—but in a play, for example, the microphones must be hidden so as not to spoil the illusion. Boom operators swing the microphones about on the ends of long booms or poles so that they are close to the actors to pick up what they are saying, but out of view of the cameras.

One important member of the studio team is the floor manager. He is also linked to the control box through headphones, and his job is to give actors and others taking part their cues. He plays a key role in keeping things to time. By gestures or holding up cards he can tell the performers how much time they have left, and can urge an interviewer to hurry someone up or even cut him short.

The control room of a television studio is usually placed so that the producer, director and others involved can look down on the studio and its activities. The director is in

charge of the actual broadcast. He has a bank of monitor screens in front of him, each one showing the picture from a different camera, or from a piece of film or tape. By pressing buttons he or one of his assistants can switch from one camera to another. A red light on the top of each camera glows to show that its picture is the one being transmitted.

A great many programmes are recorded on film or tape some hours or even days before they are transmitted. This enables the director to have the film or tape edited: he can cut a programme which is too long, or rearrange the order of events, or even reshoot a scene which went wrong and cut the new version into the programme.

Some programmes, radio and television, are outside broadcasts: that is, the microphones and cameras go to the place where an event is taking place, such as a football match or an important ceremony.

Above: In contrast with radio, the control room of a TV news studio, with a bank of monitor cameras on the right. Through the window you can see the studio.

A great deal of television is filmed on location. Here the production is *The Last of the Mohicans*.

The scene in a TV studio during the broadcast of a major drama—in this case George Bernard Shaw's *Heartbreak House*.

Sounds and Pictures

Most people like beautiful things—clothes, furniture, wallpaper, houses to live in, attractive books to read and so on. Making beautiful things is what we call the Arts. The difference between arts and crafts is that the artist is concerned with the beauty of the final product, while the craftsman is concerned with the way it is made and the skills needed to make it.

You may well feel daunted by the term 'Arts'. It probably makes you think of museums and art galleries, and people being very serious about what they are doing. Art—of all kinds—is not really like that. It is an exciting, ongoing activity, in which everyone can take part.

The arts involve things you hear and things

you see, and sometimes combinations of both. Music is something you hear and a story is something you see—you read it. Put them together, and you get an opera, something you both see and hear. Most television programmes involve both seeing and hearing, while radio is something you only hear.

People use the term 'visual arts' for things you see. They include paintings, sculptures and photographs. If art is employed in the design of a useful article, it is called 'applied art'. You could call designing furniture or buildings an applied art. If you apply art to making a useful article look more beautiful, then you are practising one of the decorative arts. Flower arranging, dressmaking and hairdressing are some of the arts under this

heading. If you are just creating a beautiful object for its own sake, such as a painting or a piece of music, then you may be said to be involved in one of the fine arts.

It is a sad fact that we are not all born with the same abilities. Some people can draw, others cannot. Some can write music easily, others can barely whistle a tune correctly. Some people have a small amount of talent, others have a lot, and it is the ones with a great deal of natural talent who in the long run do best in the arts. Even then, most of them need to do a lot of work to gain the necessary skills, so that they can make the most of their natural abilities.

However, even if you are not somebody with great artistic gifts, you can still enjoy the arts, and you can also take part in them. It is a matter of finding out what you can do, and then doing it as well as you can. Even then, it

Left: A young artist in the painting class at school. It's never too soon—or too late—to try your hand.

82

is a good idea to try something you are not particularly good at, such as painting a picture. In that way you can understand better how the artist works and gets his effects, and you will find that you understand and enjoy the work of other people much better. It is also a good idea to watch artists at work, read about them and study the techniques they use to achieve their results.

If your own efforts actually produce results, then you have the joy of having created something, and that is what art is all about. Taking part in a school play or forming your own pop group or writing a poem or a short story—these are all forms of art you can try out and enjoy.

One of the best ways to learn about art is to see and hear it. If you have a chance to go to a picture gallery, then do so. Do not make the mistake of trying to see everything at once:

you can get mental indigestion that way, just as you can get a bad stomach-ache by trying to eat too much! Select one or two pictures that take your fancy, and spend a little while looking at each of them. Notice how the artist draws or sketches and the colours he uses. In the same way, if you have a record player you should listen several times to a piece of music. Each time you will notice something you have not heard before.

In the next 32 pages you can read about some of the most important arts and learn how you can take part in them yourself. You can also see how the experts do their work in these arts. You can learn how to make music, and how music is recorded—again, something you can try for yourself with a tape-recorder. You can learn about acting plays, taking photographs and films, making pottery and sculptures and painting pictures.

School plays give every child a chance to take part in the creation of drama. In this way you can learn much more about acting and the theatre than you can by just going to watch plays or reading them.

Making Music

Music can give a tremendous thrill, both to people who make music as well as those who listen to it. You can start to make music right away. The kind of music you will make will probably depend on whether you have any musical instruments at home, or on whether you can learn music at school or join a local band or orchestra.

Always enjoy music making, and make the kind of music that *you* want to play. Also listen to all kinds of music and try playing in different ways. Classical music, popular music, folk music and all other kinds of music are all as good as each other. Find out which kind you like, and learn to play it as well as possible. Then you will get the greatest enjoyment from music, and so will your listeners.

You do not need a teacher to be able to make music. However, you will probably learn to play faster and better if you do have a music teacher. If you become good at music, you could make a living at it. If you want to be a full-time musician, then you must choose your instrument and style of music with care. Music is also a good hobby, and here it also helps to think of which instrument and which style to play in order to make the most of your talents.

The following notes on instruments and styles of music will help you if you are not sure which instrument to take up and what kind of music to play. Instruments are divided into keyboard, strings, wind, percussion and electric instruments.

The most useful keyboard instrument is the piano. Many people learn to play the piano because there is one at home or at school. The piano is a good instrument to learn because it makes splendid music on its own and yet it can also be played with other people. Furthermore, it can be played in all styles of music, and you can learn a lot about how music is put together by playing the piano. However, you will not be able to play with an orchestra unless you become good enough to perform the solo part in a concerto.

The organ is another important keyboard instrument. There are pipe organs in churches, and you can get small reed organs or electronic organs to play at home. The organ is usually played on its own. It is quite hard to play because you have to hold your fingers down on the keys, and you may also have to use your feet to play deep notes. However, the organ has the widest range of sound of all instruments (except the synthesiser), and playing a large organ is like having an orchestra at your fingertips.

The main string instruments are the mem-

bers of the violin family—the violin, viola, cello and double bass. Their sound is produced with a bow. They are played more with other instruments than on their own. These string instruments are important in classical music, and symphony orchestras have large groups of string players. You can also play in smaller groups and get more of your own enjoyment from playing. The violin is also important in folk music, and the double bass is played in popular music (many double bass players also play the bass guitar). String instruments are hard to play well, but produce music of great expression.

The guitar is a plucked string instrument. It is easy to strum chords on the guitar to accompany singing, but it is hard to play well on its own. The electric guitar is very important in popular music.

There are two main groups of wind instruments. Woodwind instruments—mostly made of wood, though some are made of metal —include the flute, clarinet, oboe, bassoon and saxophone. The first four are mainly played in orchestras, and the clarinet and saxophone are played in popular music, especially jazz. The recorder is a woodwind instru-

Percussion bands are very popular in schools. This one has a dual purpose: the teacher uses the rhythm to help youngsters with their French pronunciation.

Right: Big cello—small player; but it's never too soon to start finding out about instruments and how to play them.

Youth orchestras are gaining popularity the world over. These young Australian musicians were rehearsing for a tour of the United States.

ment played in small groups. It is good to start on the recorder as it is quite easy to play.

The other wind group consists of brass instruments—the trumpet, trombone, horn and tuba. They are played in orchestras and also in brass bands. The trumpet and trombone are important in jazz.

It is quite easy to get a good sound on a

Massed bands playing on the steps of the Sydney Opera House in Australia.

One trombone is quite enough for this young musician, let alone 76 of them. The slide of a trombone helps to change the pitch.

woodwind instrument, but a little harder on a brass instrument. Woodwind sounds are beautiful though usually rather quiet. Brass instruments can be loud and thrilling to hear.

Many children have the chance to play percussion instruments such as tambourines at school. If you are good at them you can go on to play the drums. These instruments are played in orchestras, and drums are very important in all kinds of popular music.

Today's popular music, especially rock music, depends heavily on electric instruments such as the electric guitar, bass guitar, electronic organ and synthesiser. They all need amplifiers and loudspeakers to make a sound, and they are expensive. They are played in the same way as string and keyboard instruments, although the controls can alter the sound produced. The synthesiser opens up the whole new field of electronic music, especially if you have a tape recorder. The other electric instruments are not much used outside popular music.

Of course, you do not need an instrument to make music; you can use your own voice. You can sing in front of a band or accompany yourself on the piano or guitar, or you can join a choir or a small group.

Musicians at Work

People enjoy music for many reasons. Some of us like to sit down and listen to music intently, delighting in the flow of notes and patterns of sounds. Others get up and like to dance to music, feeling its rhythms. Music affects the way we feel. A hymn tune aids worship in church; a fanfare adds importance to a ceremony; a march gets soldiers into line; and background music produces a calm atmosphere in which people can work more easily.

Music probably began in prehistoric times as people found that chanting or singing helped them to express themselves more strongly than talking. Chants and songs began to have religious purposes, and helped people to work together. Prehistoric people probably also sang for pleasure. From about 25,000 BC they began to make instruments from natural materials. They constructed flutes and whistles from bones, hollowed out horns to make trumpets, and made chimes from stones and pipes from reeds. The harp developed from the archer's bow.

Most parts of the world have their own kind of folk music. This is a traditional sort of music that is played naturally by ordinary people. They learn it as they grow up and pass it on to their children. The music does not change in style, but remains the same from one century to another.

In Asia musicians often make up the music as they play. None of the music is written down, but the musicians have a few simple rules so that they all play in the same style. However, this music is exciting to hear and play because no one is quite sure what is going to happen. In Africa the folk music is not written down either. It is full of exciting rhythms, and is often sung and played on drums. Dancing is very important. European folk music consists of songs and dances that are now written down and played much the same way every time.

In Europe another kind of music has developed over the past 1,500 years. It is often called serious music. This music is written down, and people listen to it rather than take part, unlike folk music.

This music began in the Christian Church, and at first consisted of simple chants. However, because it was so simple more complex music was able to evolve later. The chants became long musical lines, and people then started to sing separate lines together. Different musical forms developed with various sections of music to accompany church services. In this way composers began to write music to be performed by singers and then

Ludwig van Beethoven (1770-1827) was one of the world's greatest composers. He began the so-called 'Romantic' movement.

Sir Edward Elgar (1857-1934) was an English composer who came at the end of the Romantic period.

Tim Rice (above) and Andrew Lloyd Webber (below) who collaborated on the musicals *Jesus Christ Superstar* and *Evita*.

Composers at Work

Most music is made up of melody (a tune or tunes), harmony (notes that sound well when played together) and rhythm, the 'beat' of the music.

The composer writes down his tune in the form of notes (named by the first seven letters of the alphabet) on a set of lines called a stave. The position of the notes tells the singer or player the pitch (how high or low) and length of each sound. Other signs show the rests between the sounds and whether they should be fast or slow, loud or soft.

When he has written down the tune and all its accompanying parts, the composer must score them—that is, indicate which part of the music should be played by each instrument.

All this can take a long time, though some composers are quicker than others. Edward Elgar took 15 months to write his first symphony, but Wolfgang Amadeus Mozart wrote his last three symphonies in just six weeks.

also by instrumental musicians. Their music was heard in places other than churches, and was able to develop more widely.

It was not until about 1600 that the first orchestras were formed. With the development of orchestral music came opera.

The main kinds of orchestral music are the vigorous baroque music of the 1600s, which consists of several separate lines of music played together; the classical music of the 1700s, in which such forms as the symphony and concerto were developed; and the romantic music of the 1800s, in which the music is intended to create a mood or tell a story. In the present century composers have been particularly interested in extending the kinds of sounds that an orchestra can produce.

The gramophone and radio have brought music to many people who would otherwise have heard little more than their own folk music. They have come to enjoy classical music, and also folk music from other places. Most popular of all have been musical shows and songs with their pretty tunes, and jazz and rock music with their pounding rhythms. Most of this music has come from America, where people from Europe and Africa have come together and exchanged musical ideas.

Recording Sound

Just as radio waves have conquered distance, sound recording has conquered time. You can listen to the great voices and musicians of the past, or make recordings of yourself today for hearing again at any later time. In the century since Thomas A. Edison invented his phonograph in 1877, sound recorders have become simple everyday machines, and cassette tape recorders are to be found in very many homes, schools and offices.

We can record speech, music, railway trains, in fact anything that sends out through the air the waves of energy which we call sound waves. This energy makes the air pressure swing up and down from its normal atmospheric value, at a fast rate for high notes and a slower rate for low notes. You hear sounds because your eardrums take up this movement and send signals to your brain.

Recording sounds is done by two stages of energy change. The first stage is the job of the microphone. This has a thin diaphragm like your eardrum, and its shaking movements when sound waves come along are made to produce an electric current. This too keeps changing its direction of flow in step with the rate of air pressure swings. So it really is a copy of the sounds, but in electrical form. For the second stage in recording, this current must in its turn be changed into a fixed pattern.

In magnetic recording this is done by making a special tape with a coating of tiny magnets which pass continuously over an electromagnet called the record head. The current flows through the head so that as each tiny magnet on the tape passes by, it lines itself up to match the state of the current at that exact moment of time. The tape therefore carries along its length a record of the original sound wave pattern.

To play back a recording, the tape is simply wound back to the beginning and then set running again at the original speed. This time the head acts as a collector instead of a recorder; the magnetic pattern on the tape sets up a current flow in the head and this is sent to a loudspeaker or to headphones. These devices are like a microphone in reverse. They too have a thin diaphragm, and this is set into vibration when the current comes along and so sends out waves into the air for you to hear.

Recording singers and orchestra in a large studio. The surfaces of the walls have been treated to provide the best possible conditions for sound—not too 'dead' and with not too much reverberation. The acoustic screens help to reflect the sound waves. Taking part in this picture: The Ambrosian Singers, the London Symphony Orchestra, and conductor Carlos Chavez.

Recording soloists in an opera, Gaetano Donizetti's *L'Elisir d'Amore* (The Love Potion). The singers are, from left to right, Placido Domingo, Ileana Cotrubas, Lillian Watson and Ingvar Wixell.

Home recorders usually have the tape wound inside a small cassette, so that it winds from one tiny hub to another as you record or play back. The tape machines in professional studios have large reels running at higher speeds to give better sound quality. Instead of just one or two tracks, they may have up to 24 tracks on 2-in-wide tape. The engineers often use a lot of microphones at a time, and so they have complicated mixer desks where they can control the various currents. The desks have special circuits which can boost high or low notes, or add echoes to the sound to imitate the noise you hear in a large church.

When the final recorded tape is going to be copied on to thousands of gramophone records or musicassettes, it is obviously very important to get every note both performed and recorded perfectly. So the musicians may go over parts of the music many times. Then lengths of tape have to be carefully cut and joined together to produce the final version.

Sound recording is not only done magnetically. The recorded pattern can be cut as a wiggly groove on a soft disc for making gramophone records, or formed as light and dark patches on photographic film for the cinema. The latest development is digital recording, in which sounds are recorded not in a wave pattern but as coded pulses. These pulses can be stored in a computer and played back without any of the background noises and other troubles which have so far affected ordinary recording methods.

Since a tape recorder is able to record any suitable electrical signal, you can plug in a radio set instead of the microphone to record programmes, perhaps to learn foreign languages or just enjoy musical items over and over again. It is also possible to plug in a second recorder and play a duet with an earlier recording.

For outdoor recording, a recorder that runs from batteries is best. Special microphones can be used which look either like a rifle or a large saucer. They can pick up birdsong, for instance, from hundreds of metres away.

Recording engineers are responsible for placing the microphones so as to reproduce the true sound of an orchestra. Here conductor Leonard Bernstein and some of the performers listen with the engineers to a playback of a work they have just recorded.

Acting a Play

Acting on a stage is an extension of the games children play when very young, such as mothers and fathers or cat and mice. The actors each play a rôle, doing and saying what they imagine a certain character would do in their situation. This is called improvisation, and professional actors improvise to help them to understand the rôles they are playing.

You can act out many stories with a few simple costumes and 'props' (properties) such as a crown for a king or a gun for a soldier. Once all those taking part know the outline of the story, the characters make up the speeches as they go along.

If you want to act out a play on a stage with a script for the characters, you must first cast the play by deciding who is to act each part, and then begin rehearsals. Work out the movements (the technical term is 'block in') and go through the action several times. Now you are ready to learn your lines. Remember to learn the cues—the other characters' words which come just before your own. Next you will need a prompter to remind you of your words if you forget them. Actors refer to this lapse of memory as 'drying'.

Three basic acting techniques will help you: voice control, positioning and movement. Project your voice clearly but without shouting. Avoid standing directly behind another actor or you will be 'masked' from the audience. Where possible, speak to another character downstage (nearer the audience) so that you can be heard. Be bold but not too dramatic in your movements.

The director is responsible for pointing out these techniques. The term producer now means the person who provides the money to stage a play. A good director makes the dramatist's intentions clear, or interprets the play in a fresh way so that it comes alive for the audience.

While a professional actor is rehearsing, many other people are working on the presentation of the play. The designer draws up plans of the scenery and costumes. The workshop then builds and paints the set, so called because it shows the setting—the place and time of the play. The wardrobe mistress measures the actors and makes the costumes.

The set often consists of a scenic backcloth, large frames or 'flats' covered in painted canvas, and furniture. Sometimes a fly is used. This is a backcloth or other item which is 'flown' above the stage on cables, out of sight of the audience, and lowered when it is required. The props are stored in the wings, the sides of the stage invisible to the audience.

Nearer the time of the first night, the

Above: Children from 12 Australian schools, with guidance from professional actors, creating their own play around a given story-line.

Left: Last-minute adjustment to make-up on this young performer.

lighting manager starts his work. The commonest light in use today is the reflector lamp which has a wide angle, or can be focused to give a concentrated beam of light. Each light has several colour slides, usually electronically controlled, used singly or in combination to get the desired tint. Spotlights which follow the actor are now largely out of date except in pantomime or circus, and footlights have been replaced by more efficient side lighting.

Flood battens (banks of lamps attached to wall and ceiling fixtures) light the flats and

Top right: Looking down from the lighting console at the stage of the Chichester Theatre in Sussex. This stage is 'in the round'—the audience sit on all sides of it.

Far right: The lights can be adjusted from catwalks high above the stage.

90

backcloths. Most have dimmer switches to vary the intensity of the light. All stage lighting is controlled from a lighting gantry above the stage or, in very modern theatres, from a control bay 'front of house'—in the auditorium itself.

A cyclorama, which is a plain wall or large sheet of canvas stretched across the back of the stage, can be used to produce spectacular lighting effects such as a moonlit night, moving clouds, shadows and fire. Whole pictures can be cast on the screen using a projector behind it.

Another piece of stage machinery which creates a dramatic effect is the revolve, where the whole stage can turn a full circle. Often, however, this is only used to reduce the time taken to change scenes. The stage manager, who is responsible for the mechanical running of the play backstage, organises the change of scene while the actors are playing on the front of the revolve. Also backstage are people making sound effects, such as cars arriving or thunder.

Make-up is used for the final dress rehearsal. This is a skilled field. Many amateurs overdo their make-up to begin with. Current techniques can transform a young actor into an old man with authentic-looking wrinkles and bald pate. Whole false sections of face are sometimes used.

Drama through the Ages

Early drama was a development from the ritual of religious ceremonies. The Greeks were the first to have a special building for drama. The term 'theatre' comes from the Greek word meaning a place for seeing, which is a good description of a Greek amphitheatre. Tiers of seats shaped in a horseshoe were laid out on a hillside, overlooking an 'orchestra' or dancing place. Behind the orchestra was a platform stage where the main action was played, while a chorus of actors in the orchestra spoke a commentary on what was happening.

The actors wore sumptuous clothes and heavy masks for both tragedy and comedy. The gods appeared suspended from a crane above, while creatures from the underworld entered from beneath the stage through trapdoors. The great tragic dramatists were Aeschylus, Sophocles and Euripides, and their plays are still acted today.

The Romans imitated the Greeks, only making the settings more elaborate and roofing their theatres. Then the art of drama went into a decline. During the Middle Ages there were plays at fairs called morality plays, because they were like a sermon, with the actors representing general figures such as the Devil or mankind.

Scholars of the Italian Renaissance rediscovered the details of classical theatre. The Teatro Olimpico at Vicenza, built in 1585, is a beautiful example of a theatre constructed according to the ideas of the Roman architect Vitruvius, whose manuscript on architecture was published in 1486. Later theatres added the proscenium arch, dividing the audience from the stage, and painted backcloths created dramatic perspectives.

These new ideas led to a golden age of drama in Spain. There, religious plays were as popular as comedies about ordinary life. The work of Lope De Vega, Miguel de Cervantes and Pedro Calderón is richly imaginative and includes the whole of society in its scope. England also took up the new ideas and, with the plays of William Shakespeare, achieved its greatest period of drama. The design of the London theatres was a development from the courtyard of an inn, where galleries surrounded the yard. The stage projected into the audience, and the nobles were given seats on the stage itself.

Renaissance ideas spread to France, too, and flourished there. By 1650 the French drama had flowered with the great tragedies of Jean Racine and Pierre Corneille, and the comedies of Molière (whose real name was Jean Poquelin). In 1680 a group was formed which was later to become the *Comédie Française*. It is still a great theatre company today.

The imitation of classical models, called

Theatres in ancient Greece were out of doors. The audience sat on benches in a great semicircle looking down to the 'orchestra' or dancing area where the action took place.

neoclassicism, continued until the Romantic period of the late 1700s and early 1800s, when the German *Sturm und Drang* (Storm and Stress) movement produced heroic and historical plays to arouse personal emotions. Johann von Goethe and Friedrich Schiller were its chief dramatists. By 1850 drama in Europe had declined, and popular theatre was often crude melodrama. As a reaction against this a movement towards naturalism began. In Scandinavia Henrik Ibsen and August Strindberg set out to write plays about actual problems in contemporary society, as did Anton Chekhov in Russia.

In Germany in the 1930s Bertolt Brecht created a new kind of drama. He wanted his

Theatre through the ages: Top left, medieval plays were acted on simple stages in the open air; top right, Elizabethan theatres were based on the courtyard of an inn of the time; bottom left, in the later 1600s a stage had a proscenium arch, but projected forward as well; bottom right, in the 1800s the stage was completely behind the proscenium arch, like a picture in a frame.

plays and productions to disturb people and make them ask questions, appealing to the mind more than to the emotions.

The distinctive contribution of the United States to drama in this century has been the musical play. *Porgy and Bess, Oklahoma!* and *West Side Story* are examples of this form.

The types of stage now in use indicate the range of modern theatre. There is the traditional stage behind a proscenium arch. Then there is the 'thrust' stage, where the platform juts out into the audience, who are grouped around three sides. Theatre-in-the-round, as its name suggests, has the audience encircling the actors. There has also been a revival of open-air drama.

Taking Photographs

Most people like making pictures, but not everyone has the natural ability to paint or draw. These days, whether you have that ability or not you can still make good pictures by photography. Even if you do not have a great deal of technical skill, there are cameras and film that take all the difficulty out of photography. It is then up to you to exercise the artistic skill that everyone has in some degree by your choice of subject, and the way in which you select what part of it to photograph to obtain the best effect.

The most important parts of a camera are its lens and its shutter. The lens is made of one or more pieces of curved glass, and it concentrates the rays of light that shine from the subject of your picture, and focuses them on to the film—that is, makes a sharp image on the film. The film is a strip of plastic which is coated with a substance sensitive to light. It is wound from one spool to another, one picture at a time.

The shutter is a mechanical screen that normally prevents light falling on the film. When operated the shutter opens to allow the image formed by the lens to fall on the film, usually for a small fraction of a second. When the exposure has been made the shutter blocks out the light once more. After you have wound on, another image can then be recorded in the same way on the next section of film.

No image can be seen on an exposed film until it has been processed. This involves soaking the film in solutions of chemicals. In ordinary black-and-white photography, processing is extremely simple. A solution of developer causes visible images to form, and light-sensitive chemicals remaining in the film are then removed by soaking it in a fixer.

On the processed film bright areas of the subjects appear dark, and dark areas of the subjects appear light. Prints are made from these negative images by projecting them on to light-sensitive paper, which is then developed and fixed. Light values are reversed once more during this process, so that the final prints are in positive (normal) form.

Colour prints are ordinarily obtained by using colour negative film. The printing process is more complicated than for black-and-white photographs, and few amateurs do their own colour processing. Colour transparencies are usually produced from reversal film. This is processed in a special way, so that the negative images formed in the initial development process are reversed to produce a positive image on the same film.

Instant-print cameras, such as the Polaroid

Various types of cameras are shown here. A is a simple kind with a small separate viewfinder, so the photographer sees very nearly the same picture as the lens does. B is a single-lens reflex camera in which the photographer sees exactly what the camera does. C is a twin-lens reflex camera, in which the photographer sees a large picture, but through a separate lens.

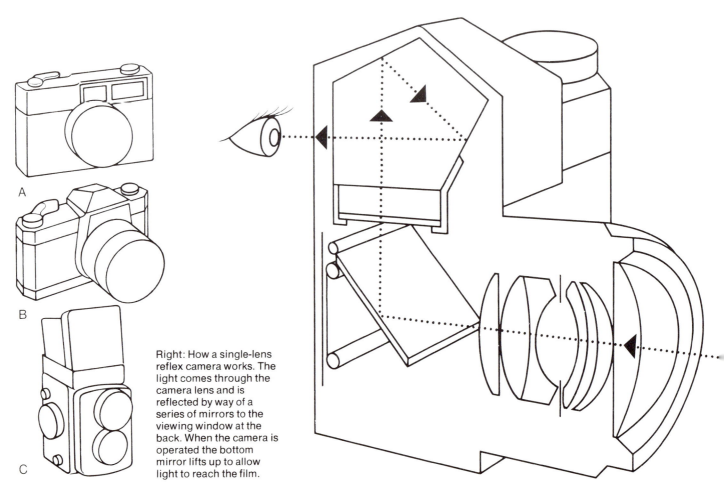

A

B

Right: How a single-lens reflex camera works. The light comes through the camera lens and is reflected by way of a series of mirrors to the viewing window at the back. When the camera is operated the bottom mirror lifts up to allow light to reach the film.

C

Left: A photograph as you normally see it—what photographers call a positive image, with the blacks and whites correct.

Right: The negative image taken by the camera, with blacks and whites reversed. The positive print is made from this negative.

models, deliver colour or black-and-white prints that develop by themselves within a minute or two. In addition, a limited range of film is available for producing instant black-and-white negatives, as well as prints, so that extra copies can be made in the usual way.

When you take photographs you must make sure, as far as possible, that the film is correctly exposed. Over-exposure—too much light falling on the film—can result in loss of detail in the highlights, or bright areas, of the picture. Under-exposure—too little light reaching the film—may cause loss of detail in the dark, shadow areas. If the camera has no provision for adjusting the exposure you must select a film suitable for the conditions in which the pictures are to be taken. Dull conditions require a fast (highly sensitive) film, whereas you should use a slow (less sensitive) film in bright sunlight.

On most cameras you can control the exposure in two ways. You can alter the shutter

speed to vary the duration of exposure, and you can adjust the size of the lens opening, or aperture, to vary the brightness of the image formed. Scales on the camera normally show the shutter speed and aperture settings selected. A scale marked 30 60 125 250 would show that the shutter speed could be varied from 1/30 to 1/250 second. A setting marked T (for Time) is often provided to enable the shutter to be held open for much longer exposures.

The aperture of a lens is usually expressed in 'f-numbers', a large f-number corresponding to a small aperture, and vice versa. A scale marked 2.8 4 5.6 8 11 16 22 would show that the full aperture of the lens is f/2.8, and that this can be reduced, or 'stopped down', to smaller apertures, the minimum being f/22. A mechanical device called a diaphragm—a circular arrangement of thin metal plates—is normally used to reduce the aperture to the required setting. On both the shutter-speed

Left: This picture has been over-exposed: too much light has fallen on the film and everything appears too pale.

Right: This picture has been under-exposed: not enough light has fallen on the film, and the scene appears far too dark.

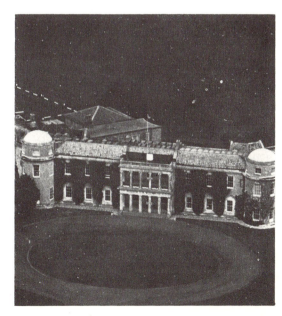

and aperture scales, going from any number to the next highest halves the exposure given.

The lighting conditions and the speed of the film in use determine which combinations of shutter and aperture settings will result in correct exposure. Instruction leaflets supplied with films give a guide to exposure under various lighting conditions, but you can obtain greater accuracy by using a light meter (exposure meter). This instrument is usually pointed towards the subject to be photographed in order to measure the light reflected from it. From the meter reading you can quickly find suitable combinations of shutter and aperture settings, using calculating dials attached to the case of the meter.

Many cameras have built-in light meters, and sometimes the metering circuit automatically adjusts the camera to ensure correct exposure. In some cameras you select a shutter speed, and the metering circuit adjusts the aperture. In others the aperture is set manually and the circuit alters the shutter speed.

Whatever aids you use to determine correct exposure, on all but the simplest cameras you still have to decide how to obtain the visual effect you want. In its simplest terms, this problem boils down to using either a large aperture and high shutter speed, or a small aperture and slow shutter speed. Both give the same exposure, but can result in entirely different visual effects.

When you press the shutter button to take a picture, any movement of the camera causes blurring of the image formed on the film. This camera-shake problem becomes particularly noticeable if the image is greatly enlarged during printing. You can eliminate camera-shake by avoiding slow shutter speeds unless the camera is mounted on a sturdy tripod, or held absolutely steady in some other way. In

Far left: In the early days of photography exposures lasted several minutes. The sitter had to keep quite still, and a rest was provided to support his head.

Left: Photography today is much more informal. The two umbrella-like shades help to diffuse the light and avoid shadows.

Right: Pictures showing the same view taken from the same place with three different kinds of lenses. Top, with an ordinary lens; top right, with a telephoto lens, which brings the image much closer; lower right, with a wide-angle lens which takes in much more of the scene.

Above left: This picture is out of focus. The shutter speed was set too slow for the speed of the horse's movement.

Left: A similar photograph, but this time the shutter speed has been increased so that the horse and rider are in focus.

practice, with a typical camera and lens, camera-shake is rarely a problem on hand-held shots if you use a shutter speed of 1/60 second or shorter.

Another factor that may influence the choice of shutter speed is subject movement. For example, you may need a speed of 1/1,000 second in order to prevent blurring when photographing a fast-moving vehicle. A slow shutter speed is sometimes used with moving subjects in order to produce deliberate blurring and thus convey a feeling of movement.

On many occasions the shutter speed is not critical, and another factor becomes more important—namely, depth of field. This term refers to the range of distances from the camera at which subjects form sharp images. A small aperture (large f-number) gives a large depth of field. This is desirable when you want both the foreground and background to be sharp. A large aperture gives a small depth of field, so that you can select, by suitably focusing the camera, which part of the picture is sharp and which part is blurred. This technique can be used to render a distracting background out of focus, so that foreground subjects stand out sharply. Most cameras have a scale that gives a rough indication of the depth of field at any aperture and distance settings.

To sum up, when shutter speed is critical, you set this first, and adjust the aperture for correct exposure. If depth of field is more important, then you choose the aperture setting first, and adjust the exposure by altering the shutter speed.

Instead of trying to master all the camera controls straight away, you can begin by adopting the following simple approach. Load the camera with medium-speed black-and-white film, set the aperture to f/16 and the

shutter speed to that recommended in the instructions with the film. For average daylight conditions, this will be around 1/60 second. Then, with the focus set to about 5 metres (16 feet), you can use the camera for general-purpose photography without any further adjustment. Just point, shoot and then wind on, ready for the next shot. Moderate changes in the lighting will not matter, because reasonable prints can be obtained from slightly under-exposed or over-exposed black-and-white negatives. The small aperture will ensure that everything except close subjects is in focus. From this simple beginning you can then progress, at your own pace, to using the various adjustments provided on the camera.

A good way to learn how to get the results you want is to study photographs that you like in magazines or books. Observe how each picture is composed, and note the camera position and the direction of the lighting.

Making Movies

Amateur cine photography, once a pastime pursued only by the rich, is now enjoyed by millions. You do not need to spend a lot at first on equipment. A camera, projector and screen suitable for the beginner cost about the same as a medium-priced 35 mm still camera. With careful planning, you can keep running costs down, so that a short film record of a holiday trip can be made for little more than a still photographer would normally spend on colour film and prints.

Whatever type of movie is being made, you should concentrate on using simple techniques

Below, upper picture: A typical modern projector for amateur use, providing sound on film facilities. Bottom: Using a splicer to edit a film.

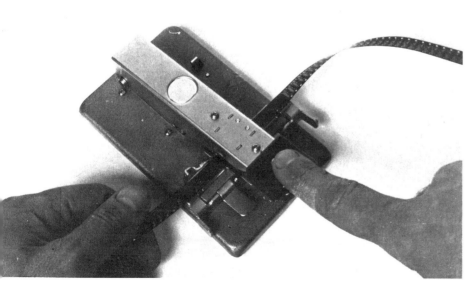

at first. Complicated shots may need to be taken several times before the required effect is obtained, and this sends costs soaring.

The most popular cine camera for amateur use takes 8-mm-wide film known as super 8. Introduced in 1965, this film has largely replaced the earlier standard 8 type. Although the same width as standard 8, super 8 film has a larger picture area. You do not need to enlarge it so much to obtain a given picture size on the screen. As a result, image quality is noticeably better. Also, unlike standard 8 film, super 8 is sold in cassettes. These are much simpler to load into the camera than the spools on which standard 8 is supplied.

Some amateur cine enthusiasts eventually progress to using 16 mm film. With this, image quality is even better, but the much higher cost of film and equipment puts it out of the range of most individuals.

The principle of cine photography is extremely simple. Numerous still pictures of a moving subject are taken in quick succession. When these pictures are later presented to the eye at the same rapid rate, the viewer is unable to distinguish the individual frames (pictures). Instead, the brain interprets the sequence as a continuously moving image.

A super 8 camera is normally run at 18 fps (frames per second). At this speed a 15 metre (50 ft) cassette lasts for 3 minutes 20 seconds. The film is alternately exposed and then quickly jerked to bring a fresh frame into position, ready for the next exposure. Usually the shutter speed is fixed at 1/40 second, exposure being controlled by adjustment of the lens aperture. In most cameras an electronic system automatically measures the light in the scene and sets the required aperture, thus making the cameraman's job much easier. Simplicity of operation is further increased on cheaper cameras by having a fixed-focus lens. So long as the subject is not closer than about $2\frac{1}{2}$ metres it will be acceptably sharp. A special supplementary lens can be fitted if closer shots are required.

The normal-focal-length lens fitted to the simplest cameras is suitable for most general-purpose filming. A telephoto lens is sometimes desirable for distant subjects, and a wide-angle lens may sometimes be necessary in order to include all of the subject when filming in a confined space. So the more expensive cameras are usually fitted with a zoom lens. This allows the angle of view to be adjusted before, or even during, filming.

The most effective way to learn techniques of any kind is through your own mistakes. In movie-making this can prove to be extremely expensive, so it is preferable to learn from the mistakes of others. Ask to see your friends' home movies—most people are only too

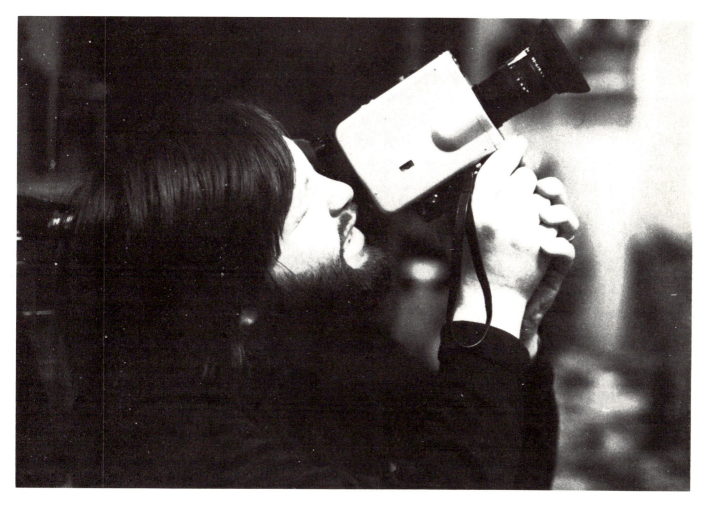

delighted to show the films they have made, no matter how poor their results. When watching the films, try to see how they could be improved.

Camera shake is a common and disturbing feature of most home movies. The camera is often moved around too much in order to follow the action. Another major fault is over-use of the zoom lens which can, in extreme cases, make the audience feel giddy.

Fortunately, these common mistakes are easy to remedy. Always keep the camera steady by using a good tripod, or else rest it on a wall or other firm support. Keep camera movements to a minimum, and turn extremely slowly when panning the camera across a landscape. If the camera has a zoom lens, try not to use the zoom effect more than once in a short film. Otherwise, the audience will become more aware of the techniques used than of the subject of the film. In general, try to ensure that most of the movement in the

Taking a cine film with a hand-held camera. Notice how the operator uses two hands to make the shot as steady as possible. The strip at the top shows how a film works: each frame has a separate shot at a different point in the athlete's movement. When the sequence is shown rapidly the shots blend into one another and the viewer thinks he is seeing continuous movement.

film is produced by the subject, rather than by the camera.

You can gain much useful information on technique by watching professional films. Look at a film on television and note how long each shot lasts, and whether it is a long shot, medium shot or close-up. Although five seconds does not sound a long time, you may be surprised to find how many shots in a film are of even shorter duration. Note also the techniques used for changing from one scene to another. For example, to break up a series of long-duration shots of zoo animals, brief close-ups of spectators might be used.

Although many techniques of the professional film maker are beyond the reach of most amateurs, superb films can still be made using only a simple camera and elementary techniques. Remember that you can always improve the film you have taken by learning how to edit it. Poor shots can be cut out, and good shots rearranged.

Film-makers at Work

Modern film-making is generally a costly and complicated affair. It involves the talents of many people, some of whom may spend several years working on one film.

When a film company obtains a story for a film—it may be a book, a play or a story specially written for the screen—it appoints a producer. He is responsible for handling all the business aspects of the film-making.

The producer often selects writers to develop the story into a screenplay, which contains the dialogue and instructions as to how each scene is to be filmed. He also often chooses the director, who takes charge of the filming. The producer and director then hire actors, cameramen, sound engineers, designers, make-up artists, editors, special effects experts and other technicians.

Film-making is not always organised in this way. Sometimes the producer is also the director, or directors may be little more than assistants to powerful producers. However, in most cases the director is the chief person in control of film-making. He often works with the writers in producing the final shooting script from which the film is made. He directs the actors during filming, and also instructs the cameramen about the angles and lighting that he requires for particular effects.

The director also works with the special effects department. For example, in a film about monsters, such as *King Kong*, the special effects department must create realistic models that appear to move.

When all the scenes and special effects have been filmed the director usually works with

the editor to link together all the shots and scenes in the best possible way. Some scenes may be cut up and arranged in a different order so as to create a more effective sequence. Therefore in many ways the editing stage is as important as the photography. Finally the sound track is prepared. The speech is mixed with background music and other sounds, such as explosions. The completed sound track is added to the film, and it is then ready for showing.

The first public film show was given in Paris in 1895, by two brothers, Louis and Auguste Lumière. They showed several short, silent films, one of which depicted a train arriving at a station. It was reported that some members of the audience were frightened, thinking that they might be run over in their seats.

Most early films were between five and thirty minutes long. For example *The Great Train Robbery*, an American film made in 1903, lasted six minutes. This film is regarded

as the first of the Westerns, a style of film that is still popular.

Gradually film companies began to make longer films, in which it was easier for them to tell complicated stories. In 1915 a three-hour film called *The Birth of a Nation* appeared. This film, made by the American director D. W. Griffith, told a story about the American Civil War. It had a large cast and great battle scenes, and Griffith tried to make every detail as realistic and accurate as possible. The success of his film led to the making of many other long story films.

In the early days many films were shot on location—that is, in real streets or other places. This was possible in such places as Hollywood, California, where it is mostly dry and sunny. However, in 1927 a major change occurred. A film called *The Jazz Singer* appeared. It had a musical soundtrack, songs and a little spoken dialogue. It was an instant success, and nearly all film companies started to make sound films. At first they faced many technical problems, especially while filming on location. So sound films were made largely in studios, huge indoor areas where the scenery is built by set designers.

In the 1930s and 1940s film-making became further complicated by the introduction of colour film. Making high-quality colour films places great responsibility on the technicians in charge of the lighting.

In the 1950s television was growing popular. To compete with it film-makers used new, complex equipment to make wide-screen films, including CinemaScope and Cinerama.

Today commercial films cost so much money to make that a film-maker's career may be ruined by one film that makes a loss.

Special trick effects were used to make this film, *King Kong*, in 1933. The city of New York is a model, while the monster (also a model) and the girl in its paw were filmed separately and combined when making the final print.

Left: Filming on location while making the mystery thriller *Murder on the Orient Express*. The express is waiting at 'Stamboul' station —actually a station near Paris, which was disguised for the occasion.

Director Irvin Kershner shouts directions to the studio team during the filming of *S*P*Y*S*.

Arts Inquiry Desk

What was the Diaghilev Ballet?
The Ballets Russes (Russian Ballet) of Sergei Diaghilev was probably the most famous of all ballet companies. Diaghilev, a Russian impresario (manager), formed it in 1909. With its male dancers, imaginative costumes and scenery and startling music, the company completely revolutionised Western ballet. The Ballets Russes created many important new ballets such as *The Fire Bird* and *Petrouchka*. The most famous male dancer was Vaslav Nijinsky. Composers who wrote music for the company included Igor Stravinsky and Claude Debussy. Right: Modern Russian ballet—a dancer in Stravinsky's ballet *Petrouchka*.

Why does a violin make a noise when it is rubbed with a bow?
If you pluck a string with your finger it makes a noise. A violinist's bow is strung with horsehair, and this hair is not smooth but is covered with tiny projections like the teeth of a saw. As the fiddler draws the bow across the strings each of the tiny projections catches the string in turn and pulls it slightly. In this way the string is receiving a lot of little plucks, one after the other, and so it sounds continuously. It is the vibration or movement of the string that causes the sound. It sets up waves in the air which eventually reach your eardrum.

? ? ?

How does a photographer make a big picture from a small negative?
The photographer uses a device called an enlarger. The enlarger has a powerful light in it, and it shines through the negative and then through a lens, which magnifies the image. The image is projected on to the table of the enlarger, and that is where the photographer lays the sensitive paper on which he makes the enlarged print. The distance of the lens from the negative can be adjusted to focus the picture, in just the same way as the camera is focused. To make the picture bigger the photographer raises the whole enlarger head so that it is further from the table. In this way he can enlarge just part of a picture, and can cut out unwanted background.

? ? ?

What is the difference between a harpsicord and a piano?
The difference lies in the way in which the strings are sounded. The harpsicord has little quills, called jacks, which pluck the strings when you press the keys. Playing the piano keys makes a series of little felt hammers hit the strings.

? ? ?

What is a sitar?
The sitar is an Indian stringed instrument. It has a deep-bowled sound box and a long neck with seven main strings which are plucked by a wire plectrum. There are 11 or 12 sympathetic strings which are not plucked but are tuned to the scale of the melody being played.

Sitars are usually very ornate; one type, the Peacock Sitar, is even carved just like a peacock. A sitar player usually sits on the floor, holding the sitar across his body. Instruments that usually accompany the sitar are the *tambura*, a type of lute plucked by hand, which plays a continuous drone, and the *tabla*, a small drum. The three instruments are used to play classical Indian music known as *ragas*. Raga means colour or feeling. Raga music is based on set rhythmic and melodic patterns and is improvised by the musicians to reflect their own mood or the mood of the audience.

? ? ?

What is an opera?
An opera is a play set to music. The word opera comes from the Italian *opera in musica* which means a musical work. Opera began in Italy during the late 1500s. At that time there was a revival of Greek drama from which opera then developed. The first great opera was *Orfeo*. It was written by the Italian composer Claudio Monteverdi, first performed in 1607, and is still

performed today. In 1637 the world's first opera house was opened in Venice. Opera spread to France and Germany. Famous operatic composers include the German Richard Wagner, the Frenchman Georges Bizet and the Italian Gaetano Donizetti whose *The Elixir of Love* is shown below left.

? ? ?

What is an oratorio?

An oratorio is a dramatic musical work for soloists, chorus and orchestra on a religious theme. It is in effect a sort of religious opera, but is performed without scenery, costumes or action. Oratorios got their name because they were first performed in the Oratory, a chapel in Rome, in the late 1500s. Later it became the custom to perform oratorios in Lent, when operas were thought to be unsuitable.

? ? ?

How did the circus originate?

The word 'circus' comes from a Latin word meaning a ring. The first circuses were held in arenas in ancient Rome. A U-shaped track was used for staging spectacular chariot races. Between races acrobats and jugglers entertained the crowds. Modern circuses stem from the late 1700s when the Englishman Philip Astley gave displays of trick horse-riding in circular sawdust-covered rings. During the 1800s travelling circuses began. Troupes of performers journeyed around village fairs and markets, giving shows.

? ? ?

What was the Globe Theatre? (*See picture on the right*).

The Globe was a round theatre, open to the sky, which was built in Southwark near London in 1598. It was at the Globe that many of the English playwright William Shakespeare's plays were first performed. Most of the audience stood in front of the stage, while the wealthier people paid to sit under cover in galleries around the walls. Plays could be watched from three sides. The stage lay above ground level and backed on to balconies, which were used in many of the plays. Behind the balconies lay the actors' dressing rooms. A flag was raised to tell people when a play was about to start.

? ? ?

How is an animated cartoon made?

A cartoon is a series of still pictures photographed by a film camera. When the film is shown on a screen all the images run together to give the impression of movement. To show a person running, for example, each position of the body is drawn separately and in sequence. As the film rolls so the person seems to be running along. One of the most famous animators was the American Walt Disney (1901-1966). His well-known character, Mickey Mouse, was first seen in the 1920s. Cartoon-making is a long process; for example it took Disney three years to make *Snow White and the Seven Dwarfs*.

? ? ?

What was the earliest type of camera?

The camera obscura was the forerunner of the modern camera. It consisted of a light-proof chamber or box. In one wall was a convex lens through which light fell on a screen to produce an image. During the 1600s the camera obscura was used mainly as an aid to artists.

The Frenchman Joseph Niepce (1765-1833) was the first person to turn the camera obscura into a photographic instrument. He used a microscope lens and replaced the screen with a chemically treated pewter plate. In 1826, using his converted camera obscura, he took the first photograph.

Making Pottery

One of the most satisfying crafts is making pottery. The work can be as simple or as complicated as you like, or as your skill allows—but in either case the result can be good. You need very few tools: wooden sticks and spatulas for shaping the clay and smoothing it, small sponges for moistening it while working, and wire tools for cutting into the clay. The one really expensive piece of apparatus is the kiln in which the pottery is fired at a very high temperature, and using a kiln requires skill and experience. Fortunately there are many chances of doing pottery at school or in evening classes, so you can have your pottery fired for you.

Clay must first be wedged (worked) to get rid of lumps and air bubbles. Small quantities are hand wedged. Form the clay into a ball and use a twisting, slicing movement of the

Left: A skilled craftsman throwing a pot on the wheel. He shapes it entirely with his hands.

thumbs to tear the ball in half. Then tear the clay in half again, turning the ball to cut through the join. You should repeat the process until the clay is smooth and evenly mixed. Larger amounts are bench wedged. In this process you slap the clay on to a bench, ease it off and then slap and bump it into an oval shape. Place a cutting wire under the clay and slice it in half. Lift one half and bring it firmly down on top of the other. This explodes the air bubbles and makes the clay stick to itself. Form the clay into shape again and turn it round to cut across the join. This should be repeated until the clay is smooth and workable. Do not dent it with the fingers because this causes air pockets.

After wedging, the clay is ready for use. Practise making small shapes. The clay will dry out as it is handled and should be kept moist with a small sponge dipped in water. A

Opposite: Wedging clay is an important preliminary to successful pottery—and it's also great fun.

pinched dish can be made from a ball of clay by pressing the centre with the thumbs. Pinch round the hole with thumb and forefinger. This will form the walls of the dish.

Coil pots are made with clay 'ropes'. The usual method is to roll out the clay on a bench by hand. The rope should be of even thickness. Flatten a ball of clay into a round for the base of the pot. The first coil is placed on the base and joined at the two ends to form a ring. You can join this ring to the base by smearing clay from the inside of the coil down on to the base. Then you can put one coil on top of another until the pot is large enough. Coils should always be joined on the inside.

Coil pots are usually rounded in shape. For flat-sided objects such as boxes or trays, slabs are used. To make slabs, roll the clay flat on to pieces of cloth, like rolling out pastry. Let the clay stiffen slightly before cutting out shapes. When making a box, you fix the joins together by dampening the edges with water. Put a thin coil along the inside edge and smear it into place, using a smooth modelling tool.

Casting is done by filling a hollow mould with slip (liquid clay). Allow the slip to set slightly, then pour out the rest of the liquid. The remaining clay sticks to the inside of the mould and takes an impression of it. Leave it to dry and it will come away from the mould.

A more skilled form of pottery is throwing on a wheel. The potter's wheel is driven either by a foot pedal or by an electric motor. It takes a lot of practice to make good thrown pots. Begin by throwing a ball of clay on to the centre of the wheel so that it sticks. As the wheel turns you put one hand into the centre of the ball to hollow out the pot. Make the wall of the pot by drawing the clay up between the hands, keeping an even thickness for strength. Cut the finished pot off the wheel with a wire and leave it to harden.

You can decorate pottery in many ways. The simplest method is to scratch or scrape a pattern into the surface, using tools or even your fingertips. You can cut holes in the walls of a pot for decoration. Make separate decorations from spare clay and stick them on the pot, using slip. You can also cast decorations from slip and stick them on.

To colour their pots, potters use metal oxides, which can be bought ready mixed. You can use them to paint designs on the pot or to provide an all-over colour. Remember that the colour of the raw oxide is not necessarily the same as it will be when it is fired. The colours are usually put on after the first firing in the kiln, when the pot is in a plain state known as biscuit. After decoration the pot is dipped in glaze, a glass-like substance, and fired again. This second firing fuses the glaze on to the pot and makes it waterproof.

Making Sculptures

There are two main ways of making a sculpture. You can build it up by modelling with such materials as clay, plaster and Plasticine, or carve it out of stone, wood, wax and even soap. Using any of these materials, you can create a whole range of objects from simple figures to complex models.

For modelling there is a special type of clay that hardens without heating. It can be painted and varnished. Start by making small animals from balls of clay. Pinch out legs, ears and noses with a finger and thumb, and mark eyes and nostrils with a pointed stick. A spider can be made with pipe-cleaners or wire pushed into the clay to form legs.

You can make small figures from clay coils. Roll out two thin coils and a third thick coil to form an H shape. Add a small ball for the head. Bend the arms and legs into position. The ends of the arms are pinched to form hands and the ends of the legs are turned up as feet. You can make clothes with strips of spare clay.

When you make larger figures and animals the legs must be firm enough to support the weight of the body. You can buy special hardeners which will strengthen the model. They are painted on before the model is coloured or varnished. If you are making an animal such as a horse, you can use string for the mane and tail. If the horse is to have a rider, let the horse dry out a little before placing the rider on top.

Clay is too heavy for larger sculptures or models. Instead you can use plaster of Paris or filling plaster. Plaster of Paris sets very quickly; filler stays soft for nearly an hour. You can make plaques or wall plates by using a shallow container, such as a round cardboard cheese box, for a mould. A loop of string placed in the box acts as a hanger. Fill the mould with plaster and make a pattern by pressing pips, seeds or buttons on the wet plaster. When dry the plaque can be painted with bright colours and varnished.

For large-scale models a framework is needed. Use modelling wire and Plasticine on a board base. Coat the framework with plaster and leave it to set before painting. A framework for a figure is made from wire wrapped with strips of pasted paper to give it body. It should be left to dry and then covered with plaster, using the fingers for smoothing.

Carving tools are sharp and can be dangerous unless you have the experience and strength to control them. Instead you can do basic carving with a small hacksaw, a rasp and a penknife. The wood or stone to be carved must be held firmly either in a vice or in the hand. If it wobbles about the blade may slip. Always carve with the blade pointing away

A carved face emerges from a block of wood. The texture of the raw wood around the face heightens the dramatic effect.

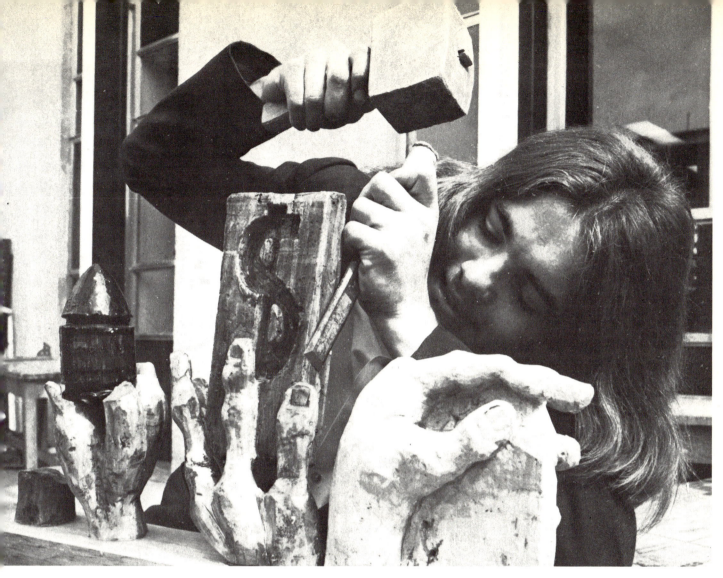

from yourself. Carve a little bit away at a time; if you try to cut off too much the knife may bend or break and cause an accident. Make sure the hand not holding the knife is behind the blade so that it will not be cut if the knife slips. This particular advice applies to all edged tools. You can avoid all risk of accident by keeping both hands behind the cutting edge.

Most types of wood can be carved. Soft wood such as balsa is light and easy to cut. First draw the design on one side of the wood and saw away the larger unused areas. The wood should be shaped using a penknife and rasp. The finished carving can be smoothed with sandpaper and then painted. Several coats of varnish will be needed because balsa absorbs moisture easily.

Plywood makes interesting sculptures as the different layers are revealed when they are carved. Varnish shows up the different colours in the wood. Blocks of wood can be used as mounts for the finished carving. Whittling is cutting shavings of wood from a stick with a knife. Bits of bark can be left on a whittled stick to make the design stand out.

Soft stone is easy to carve. A junior hacksaw can cut through limestone, sandstone and slate. Again, draw the design on the block and

A student at work in a woodcarving class. The hand is part of the carving: sculptors and woodworkers are always careful to keep their own hands behind the blade of a chisel.

cut out a rough shape with a hacksaw. Use a vice to hold the stone in place. With a rasp you can shape the stone to the drawn design. Varnish should not be used because it spoils the surface of the stone.

A simple shape can be cut from slate using a penknife. The rasp is used for shaping, and the object is finished off by smoothing with sandpaper. Coating the slate with wax will bring out the natural colour. It can then be polished with a soft cloth.

Soap can also be carved into simple figures or designs. Use a penknife to carve out the design, a little at a time. If the carving breaks, the broken edges can be softened with water and pressed hard together. The finished work can be smoothed with the knife blade and polished with a cloth. Wax candles can also be carved with a penknife. Paint can be rubbed into the candle, left to harden for a few minutes and then rubbed off. The paint that is left shows off the carved details, and the candle can then be polished.

Natural objects such as pebbles and driftwood need no carving to stand as sculptures in their own right. Pebbles can be painted or varnished; interesting pieces of driftwood can be sandpapered, varnished and then mounted on a block.

Sculptors at Work

Sculpture takes many forms. Statues, relief carvings and bronze casts are all different forms of sculpture. Artists have sculpted for thousands of years. Their basic methods have stayed much the same but the different materials and tools that they have used have led to many different styles. One of the oldest examples of sculpture dates back some 25,000 years. It is a small carved ivory figure, probably a charm used in magical rites.

Much primitive sculpture was connected with mythology and religion. Carved figures representing various gods were sculpted in animal and human form. They had large heads, protruding eyes and bared teeth to show the power of the god. Fertility gods were represented by carved pregnant female forms. The ancient civilisations of Central and South America worshipped the Sun. Their artists sculpted in gold to represent the Sun god.

The sculpture of ancient China was very decorative. Tiny, delicate figures were carved from ivory and jade. For larger objects the Chinese sculpted in bronze, using a technique called the lost wax method. In this process the figure is first modelled in clay, and a mould is made of it in plaster of Paris. This mould is lined with wax, and sand or plaster is then poured in to form an inner core. Finally the mould is heated and the wax runs out, leaving a space into which the metal is poured.

In ancient Egypt sculpture played an import-

Above left: An Egyptian statue of the sky-bearing god An-Hur. Egyptian sculptors made all their statues very formal, and standing figures usually have one foot forward.

Above right: A Roman statue, also of an Egyptian deity (Isis), shows a great contrast in styles. The figure and its pose are both lifelike, and the folds of the robes are shown in great detail.

Left: This ivory statue was carved in Japan about 1800, and in its smooth lines is typical of oriental sculpture. It shows Shou Lao, the god of long life, with a deer.

ant rôle in religious rites. Egyptian artists used a series of mathematical formulae to measure their sculptures, which they carved from hard granite. This resulted in a style which was stiff and very formal. The finished sculpture was painted to appear more life-like. Eyes were made from ivory and coloured glass or jewels, and were set into the stone eye socket. The Egyptians decorated the walls of temples and burial chambers with intricate relief carvings showing scenes from a dead person's life.

Early Greek sculpture was also painted and was mainly incorporated with architecture to decorate temples. Gods and goddesses were sculpted in human form. Instead of columns a single carved figure known as a caryatid was often used to support part of a building. Later Greek sculptors used close-grained stones such as marble to give a more flowing and natural effect. Their style is known as the classical Greek style. Greek and Roman artists also practised terracotta sculpture, in which statues were made from clay and then baked.

For some years after the end of the Roman Empire fewer sculptures were made, but from the 1200s sculpture flourished again, particularly in Italy. Sculptors of the Renaissance, the period from about 1400 to 1600, were influenced by the classical Greek style and tried to achieve near-perfection of the human form in their statues. Some of the greatest sculptors of the Renaissance included Donatello, who sculpted figures on horseback, Luca della Robbia, who perfected the art of enamelling on terracotta figures, and Michelangelo, who sculpted the famous statue of David. The statues of this period were so lifelike that they did not need to be painted.

The Baroque sculpture of the 1600s was very grand and ornate; many churches were decorated in this style.

During the late 1800s there was an upheaval in the world of art. Painters and sculptors alike tried to show realism in their work. The French artist Edgar Degas produced a beautiful sculpture called *The Little Dancer* which was complete with real shoes, clothes and hair ribbons. Again in France Honoré Daumier left marks of fingers and tools on his finished pieces to give his sculptures a striking lifelike appearance. Auguste Rodin created *The Age of Bronze*, a male figure rousing himself from sleep. It was so lifelike that Rodin was accused of having taken a cast from a living person. He created this effect by modelling the surface of the bronze to attract the light.

A little later in Italy the Futurist sculptor Umberto Boccioni tried to convey movement in his work. His best-known sculpture is *Unique Forms of Continuity in Space*, which takes the form of a striding man. The Roma-

Many sculptors today create simplified forms to convey an idea rather than to show a likeness. The English sculptor Henry Moore carved *King and Queen* in 1952-53.

nian sculptor, Constantin Brancusi, tried to break down natural form into symbols. He reduced objects to their simplest form with sculptures such as *The Beginning of the World*, which is a simple smooth egg shape based on the outline of a human head.

With the influence of abstract art, sculptors began to experiment with all sorts of different materials such as wood, glass, plastics, steel and everyday junk objects. At the same time many modern sculptors were also influenced by primitive art and sculpture. As a result modern sculpture varies a great deal—there is no one style.

In England Henry Moore expressed basic form rather than realistic likeness in his sculptures, many of which were made from wood carved with holes. Sculptors such as the Frenchman Jean Arp and the American Alexander Calder chose to work in metal, constructing mobiles—moving sculptures—and metal sculptures rather than modelling. The American artist Andy Warhol used everyday commercial objects such as soup cans to make sculptures, and other artists used wheels, chains and similar mechanical parts.

Making Pictures

Almost anything that makes a mark can be used for making a picture. Charcoal, pastel crayons, pencils and pens can all be used for drawing. The most common paints include acrylic, gouache, oil, tempera and water-colour. All of these have different qualities.

Pencils can be hard or soft. Hard pencils give fine lines. Soft pencils give broad lines and can be used for filling in and shading. Pen nibs vary in thickness to make thin lines or thicker broad lines.

Pastel crayons or chalks come in many colours. You can break them so that you can use the corners for line drawing and the long edges for shading and filling in. They are generally used on tinted paper. Chalks are dry and it is difficult to mix colours, but you can make mixtures by putting one colour on top of another. Charcoal is used in the same way as chalk, but colours are limited to black and brown.

Water-colours are thin, transparent paints which are washed on to damp paper. They are ideal for landscapes. Water is added to lighten the colours. White is rarely used in water-colours, so bright areas are left free of paint for the white of the paper to shine through.

Tempera is a paint mixed with egg-yolk. It is very quick-drying and is usually thinned with water. Gouache, also called poster paint, is also thinned with water. Poster paints are all bright colours which do not fade and can be used for fresco painting. In fresco, wet plaster is spread on a wall, and paint is put on top of it. When the plaster dries it fixes the paint.

Acrylic paints have a plastics base. They are very fast-drying, and you can put them on thinly with a brush or thickly with a palette-knife. They can also be thinned with water.

Oil paints can be put on thickly with a brush or a knife, or you can mix them with linseed oil to produce a thin glaze. Oils can be painted on any surface, although canvas is the most usual. The surface must first be primed, that is, have a coat of white priming paint of some kind before you put on the oil paint. It is best to paint with the canvas supported on an easel. Mistakes can be corrected by lifting off the paint with a palette-knife.

Before starting a painting or a drawing it is best to choose an interesting subject which is easy to reproduce. Making a subject come alive can be difficult. There are various ways of doing this. The use of shading gives a three-dimensional effect. Light tones can be produced in a drawing by making a series of lines far apart. For dark tones the lines are thick and close together. Drawing lines diagonally over the top gives very dark shadows and is

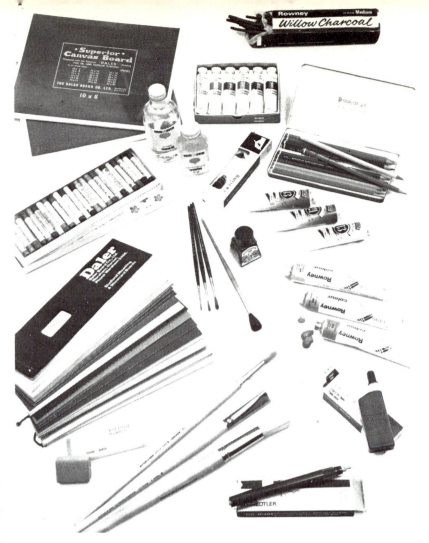

Some of the tools of the artist's trade: From the top you can see charcoal; canvas board for painting on; poster paints (gouache); bottles of linseed oil and turpentine for oil paints; pastels; drawing pen; crayons; brushes, ink; oil and acrylic paints; samples of paper; erasers; and another kind of pen with special black ink.

Opposite: A fine example of perspective is seen in this picture of the *Annunciation* (the visit by the Archangel Gabriel to Mary), painted by the Venetian artist Carlo Crivelli in 1486. Notice how the lines running from the front to the back of the building draw nearer together as they appear to get further away from the viewer.

known as cross-hatching. The same effect can be produced by using dots instead of lines, a process known as stippling.

Colour can also be used to suggest different effects. Reds and oranges suggest warmth, while blues and greens are cold colours. Using colour so that it fades into the distance creates a feeling of space between the foreground and background.

Distance or depth can be given to a picture by using perspective. This is the art of drawing objects as they really appear to the eye, when things close to us look larger than identical things further away. Perspective makes objects in a picture seem to disappear into the horizon in the way that the sides of a road seem to meet each other in the distance.

You can experiment with paint in many ways to produce different effects. It can be painted on a surface with a brush, or sprayed or sponged or put on with fingers. Pencil lines or charcoal can be smudged to produce fuzzy effects. Different textures can be produced by putting paint on thinly or thickly, or by mixing it with something such as sand.

All sorts of materials can be used to make pictures. Beads, beans, bottle tops, buttons, shells and wood can all be stuck on top of paint, or glued directly on paper. You can cut up and arrange cloth, coloured paper or newspaper to make a picture known as a collage.

OPVSCARO
LECRIVELLI
VENETI

1486

LIBERTAS · · ECCLESIASTICA ·

Artists at Work

Man has always painted pictures but styles and techniques have varied through the ages. Stone Age people painted hunting scenes on the walls of their caves using primitive tools and earth colours. The ancient Egyptians decorated their tombs with stylised flat figures, beautifully drawn.

The artists of ancient Greece were highly skilled although little of their work has survived. They used a technique known as encaustic. They mixed pigment with molten wax and put it on the surface while it was still hot. It gave rich colour effects. Medieval artists usually painted in tempera. They mixed their pigment with egg and applied it to a card or wooden panel treated with gesso, a type of plaster. The gesso shone through the painting and gave a luminous effect, very suitable for religious paintings.

Until the end of the Middle Ages most painting was influenced by religion. Byzantine artists of the Eastern Roman Empire perfected small religious pictures known as icons. Christian artists all over Europe painted beautifully illustrated manuscripts, and decorated the walls of churches with huge frescoes. During the Renaissance (the 1400s-1500s) painting styles and techniques exploded as artists experimented with perspective and colour. In Italy fresco and tempera painting reached their height. Two of the greatest Italian fresco artists were Michelangelo (1475-

This study of the Madonna and Child with a cat is one of many Leonardo da Vinci made. Artists often do sketches like this as preliminary work before a painting.

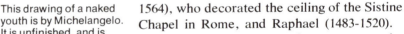

This drawing of a naked youth is by Michelangelo. It is unfinished, and is obviously an example of an artist trying out ideas.

1564), who decorated the ceiling of the Sistine Chapel in Rome, and Raphael (1483-1520).

A little earlier, in northern Europe, the Dutchman Jan van Eyck (1370-1440) was one of the first artists to perfect the use of oil paint and glaze. The development of oil paint marked a turning point in painting techniques. It gave artists a much wider means of expressing light and colour. In Italy the great Renaissance artist Leonardo da Vinci (1452-1519) used oils to paint his marvellously life-like portrait known as *Mona Lisa*.

From the 1500s canvas was increasingly used for painting. Artists found that by varying the thickness of paint or the way it was put on canvas different effects of light and shade could be produced. The Dutch artist Rembrandt van Rijn (1606-1669) used a scuffing technique, dragging light paint over dark tones to produce extreme contrasts of light and shadow. For his portraits Peter Paul Rubens (1577-1640) used a combination of oil and tempera to produce a mother-of-pearl effect very much like real skin. Thomas Gainsborough (1727-1788), one of the greatest English landscape and portrait artists, used a

One of the finest landscape artists was John Constable, and this painting of a cottage set amid fields and trees is a fine example of his work.

glaze over areas of more solid paint to produce warm and luminous effects.

During the 1700s and 1800s a movement known as Romanticism developed. Artists moved away from classical styles and began to concentrate more on atmosphere. The English landscape painter John Constable (1776-1837) reproduced the freshness of nature, capturing the effects of sunlight and clouds in his paintings. Another great English romanticist, William Turner (1775-1851) painted atmospheric landscapes.

Later in the 1800s Romanticism gave way to the French movement known as Impressionism, which revolutionised art and paved the way for all modern movements. Artists became more concerned with creating an impression of their subject than with making a detailed copy. Later still, artists such as Paul Cézanne (1839-1906) and Vincent van Gogh (1853-1890) used bright, contrasting colours in bold imaginative ways to create impressions of light and shadow.

Georges Seurat (1859-1891) introduced a technique known as pointillism. He did not mix his paints on the palette but applied them straight on the canvas in dots of pure colour to give a light, airy effect.

Since the early 1900s art has become increasingly abstract and has taken many different forms and styles. One of the leading figures was the Spanish artist Pablo Picasso (1881-1973). He pioneered many revolutionary styles, including Cubism. In this technique the subject was broken down into geometrical forms by the artist's use of flat colours and by his painting two views of a subject one on top of the other. Other major abstract artists included Paul Klee (1879-1940) and Wassily Kandinsky (1866-1944), who both used watercolours for their highly imaginative paintings.

In the 1920s artists such as Salvador Dali (born 1904) painted dream-like fantasies in a style known as Surrealism.

Today art techniques have become very flexible and there are many different styles.

Gallery of Art

Here are some outstanding examples of the work of artists from 20,000 years ago to the present day. They show just a few of the different ways in which painters can achieve their effects.

Above: This picture of a bison was painted by a Stone Age artist more than 20,000 years ago. It is on the wall of a cave at Altamira, in northern Spain. Many cave paintings of this kind have been found. People think they were made to bring good luck in hunting.

Above left: This Greek icon (religious panel) was painted by an unknown artist in the 1300s. It is typical of the severe, formal style of the times.

Left: A wall painting of 'The Flight into Egypt' is attributed to Giotto, who is known as the 'father of modern painting'. Notice the lifelike faces, although the picture was probably painted some years before the icon.

Right: Modern artists are always looking for new ways of making pictures and expressing their feelings. In 'Whaam!' and similar pictures the American artist Roy Lichtenstein created a whole new style of art in the 1950s by using the ideas of comic-strips. This kind of painting is known as 'Pop Art'.

114

Above: The English artist J. M. W. Turner painted this view of Venice in the early 1800s. In his ability to create 'atmosphere' Turner was in advance of the Impressionists of the later 1880s.

Left: The Dutch painter Rembrandt van Rijn, who lived in the 1600s, excelled both in paintings of religious subjects and in his skill at capturing faces and their expressions. Both are shown in 'The Holy Family With Angels' (1645).

The French artist Paul Gauguin painted this scene in Tahiti in 1893, when he was living in the South Pacific islands. He used bold, flat colours and simple outlines to convey his ideas.

115

People and Places

All human beings, wherever they may live, belong to the same species: *Homo sapiens* (thinking man). Hence they are all closely related; but this does not mean that they are all alike. People differ in the colour of their skin, their height, types of hair, shape of head and face and the colour of their eyes.

There are various reasons for these wide differences in shape, size and colour. One of them is the fact that many groups of people have lived apart for thousands of years. You might be forgiven, for instance, for thinking that the tall, ruddy-cheeked, fair-haired, blue-eyed Scandinavian belonged to a completely different species from that of the tiny, black Pygmy of central Africa, with his round face and dark, fuzzy hair. Such differences probably evolved over ages of time as each group adapted to its particular environment.

The tall Scandinavian would find it difficult to move silently and effortlessly through the Pygmy's forests, his very size hampering him; while at the same time he would suffer agonies from the effects of the fierce African sun on his fair skin. The Pygmy, on the other hand, would find his lack of inches an embarrassment in a big land of mountains and fjords, where muscle and bulk are needed to challenge the elements. The large amounts of melanin in his blood, which tint his skin and protect him from the sun, would be of little use to him as he cowered from the biting Arctic winds.

One of the difficulties in identifying and classifying the various races, and allocating them to certain parts of the world, is that many large natural groups have outgrown their original homes and spilled over into adjoining territories or travelled far to other parts of the Earth. Another difficulty is that many races have mixed and inter-married for hundreds of years, and the particular characteristics of the original races have become fused and blurred in the offspring.

The passing on of characteristics from parents to children—a process called heredity—is controlled by tiny specks of matter called genes. Race changes result from the degree to which certain genes are present in the blood. A person's genetic make-up can also be affected by a number of outside factors, one of which is natural selection.

Natural selection is the process by which people whose genetic make-up is right for their particular surroundings tend to survive and reproduce. Those whose genes are not right for those circumstances tend to die off before they are able to have children. The colour of the skin and the body's framework

are two characteristics that experts believe are most affected by natural selection.

The influence of the environment on the physical characteristics of populations over thousands of years has led experts to adopt a classification of races on the basis of geography. Such races exist because they have been separated by physical barriers such as oceans, deserts and mountains. American Indians, for example, were separated from the rest of the world until Christopher Columbus rediscovered America at the end of the 1400s. Similarly the Aborigines of Australia were left undisturbed on that isolated continent until British convicts began arriving there during the 1700s and occupying their lands. Not all

geographical races correspond precisely to continents. The European geographical race, for example, extends to the Middle East and South America, as well as to the white populations of Australasia.

Some experts recognise nine geographical races: European, Asian, American Indian, Eskimo, African, Australian (Aborigine), Polynesian, Melanesian and Micronesian.

There is also another side to the coin. There are many parts of the world where man has changed his environment quite as drastically as the environment has affected him. A tragic example is the so-called Dust Bowl region of the United States. In the 1930s farmers in the Great Plains area in the south-western United States removed the hedges, trees and windbreaks—which formed a protective cover of vegetation—in order to cultivate more crops. As a result during subsequent droughts the wind simply blew the fertile topsoil away and 20 million hectares of farmland were transformed into a veritable dust bowl in which nothing could live. This led to great hardship among the farming communities and mass emigrations to other, more hospitable, parts of the country.

Similarly, the water-level of the Aral and Caspian seas is dropping steadily because Russian engineers have diverted some of the rivers that feed them, using the water to irrigate farmland.

Whatever colour children's skin is, they still enjoy playing with the same things.

People of Europe

Europe is the second smallest of the six inhabited continents of the world. Only Australia is smaller. Yet Europe has more people than any other continent except Asia, the largest. More than one-sixth of all the people in the world live in Europe. These people are more crowded together in the space available than people in any other continent. There are 64 Europeans to every square kilometre of land.

On the eastern side Europe is attached to Asia. A chain of mountains, the Urals, forms the boundary. For the rest, Europe has an irregular coastline made up of countless bays and inlets. As a result the coastline stretches for 80,465 kilometres. Compare this with Africa, a continent that is three times as large but whose coastline is less than one third as long as that of Europe.

In spite of its comparatively small size

Above: A group of young people in regional costume in the Epirus region of Greece. Many of the people of this region are Tzintzars, who are related to the people of Romania, the Vlachs.

Left: Harvesting grapes in the Côte-d'Or region of Burgundy, in east-central France. Burgundy is one of the centres of France's wine-making industry.

Europe is where Western civilisation began. From the time of the ancient Greeks until the 1900s no other continent has had such an influence on world history. It has been the leader in politics, science, art and philosophy, and the civilisations of the Americas and Australasia have largely developed from it. Was it an accident that such a concentration of peoples and ideas should find themselves in Europe, or was there some reason for it?

Most people enjoy living in a temperate land—one where there are few extremes of heat and cold. Europe for the most part fulfils these conditions. The warm waters of the Gulf Stream flow from the tropics into the North Atlantic Ocean and wash Europe's long coastline. This results in mild winters and warm summers.

The winds in Europe usually blow from west to east, taking the warm air inland from

the ocean, and there is another unusual feature: the mountain ranges run from west to east. As a result they do not block the warm air that sweeps across the continent, as they would do if they ran from north to south. The winds also bring moisture and, with plenty of rain, farmers find it easier to grow crops. So these were obviously good reasons for farming communities to settle and prosper.

In spite of these common factors, some parts of the continent are very different from others. In the far north, for example, where Arctic waters freeze round parts of Scandinavia, and also in the north-east in European Russia, the winters are bitter. The peoples of those regions tend to be hardy and for the most part fair, with blue eyes. Life is a continual battle for survival. Huge forests of birch, fir, pine and spruce yield valuable timber and fuel.

The peoples of northern Europe also look to the sea for a living. Hence the Scandinavians, for example, descended from the roving Vikings, are regarded as fearless fishermen, seal-hunters and whalers. They are also experts at winter sports such as ski-ing and skating.

The great mountainous areas of Europe—the Alps in south-eastern France, northern Italy, southern Germany and most of Switzerland; the Carpathians in Hungary; the Apennines in Italy; the Pyrenees between France and Spain; and the Caucasus in Russia—have separated the various countries and allowed each to preserve its own way of life. At the same time plentiful passes in the mountains have encouraged trade and the exchange of ideas and cultures.

Dairy farms spread across some of the

Left: Europe is rich in historic remains, such as the ruins of Caerlaverock Castle in Scotland. It was captured by Edward I of England in 1300 during one of the many wars between England and Scotland.

Below: Ski-ing is a popular winter sport, especially in the Alpine regions of Europe. These skiers are speeding downhill at Pontresina, near St. Moritz in Switzerland.

lower Alpine slopes. In parts of Austria and Switzerland centuries ago the farmers found themselves with little to do during the long, dark, winter evenings, so they began to make small objects by hand indoors. Today the skills they developed have paid off in the beauty of their hand-carving and the precision of their smaller mechanical products such as clocks and watches.

A great central plain stretches from the English lowlands through northern France, Belgium, the Netherlands, northern Germany, Poland and European Russia. The soil is rich in coal and iron-ore deposits, and there many of the great mines and manufacturing centres are located. It was this natural wealth that was largely responsible for changing western civilisation from a mainly farming community into one based on manufacturing since the start of the Industrial Revolution in the 1700s. However, agriculture is important in many European countries such as Denmark and the Netherlands, which have applied industrial methods to their farming.

In the far south the climate takes its name from the Mediterranean Sea, which separates Europe from North Africa. There the summers are long, dry and hot and the winters are wet. The peoples of Portugal, Spain, southern France, Italy and Greece tend to be darker-haired and swarthier-skinned than other Europeans. Life may not be easier than in the harsher northern countries, but it is slower and more relaxed. The people grow grapes for wine-making, olives and citrus fruits, and the Mediterranean yields many kinds of fish. There is a natural tendency to sit in the sun and watch the world go by.

People of Asia

Asia, the largest of the continents, covers nearly a third of the land area of the Earth. It also holds more than half the world's people.

Scientists at one time grouped mankind into three main races: Caucasoid (white-skinned), Negroid (black-skinned), and Mongoloid (yellow-skinned). They realised that this was an over-simplification, and today many experts prefer to group people according to geographical races, of which there are nine.

According to the new system, Asians may be Asian, Indian, or European. Those living in areas to the east of India (the former Mongoloids) are now Asians. Those found in India, Pakistan and Bangladesh are Indians. Those living to the west of India are European Asians. Both Indian and European Asians belong to the old Caucasoid group.

In North Asia, which consists of an immense part of Russia, the peoples are a mixture of Asians and Europeans. Central Asia, made up of Mongolia, Sinkiang and Tibet, is the bleak home of more Asians.

There are two main areas east of India. These are East Asia, which includes most of China, Japan, North Korea, South Korea and Taiwan; and South-East Asia, which extends

Above: A street food vendor in Delhi, India's capital, prepares his wares. In a warm climate like that of India people tend to live out of doors a great deal.

Left: A girl from the Soviet republic of Tadzhikstan, which lies on the borders of China and Afghanistan. The people of this part of the Soviet Union are closely related to the Iranians.

from Burma south-eastwards as far as Singapore, Indonesia and the Philippines.

Most of the peoples of East Asia are Chinese. They are of Mongoloid stock, distinguished by yellowish skin, coarse, straight black hair, and eyes with lids that fold in the inner corners, making the eyes look slanted. The Japanese, who are not much different in appearance, come from the same stock, probably mixed with Ainu blood.

The Ainu are a very ancient people (about 15,000 survive today) who live in north-eastern Japan. They are white, short and stocky, and the men have long hair and beards. The Ainu have intermarried with the Japanese, and this may account for the fact that most Japanese have short, sturdy legs. Their comparatively small build has contributed to the development of various forms of unarmed combat (at which the Japanese have traditionally been masters) as a method of self-defence against bigger opponents.

The only Asian Negroid peoples are found in parts of Malaysia, New Guinea and the Philippines. Among them are some Pygmy races, similar to those who live in Africa.

Indian Asians can be divided into two main groups: Indo-Aryans and Dravidians. Indo-Aryans are light-skinned and live in northern India, whereas the dark-skinned Dravidians live in the south of the sub-continent. Pakis-

120

tanis are made up of these two types as well as having Arab ancestors.

Centuries of overwork, under-nourishment and ill-health among the millions of peasants have produced an Indian 'type'. This can be identified by dark hair, expressive dark eyes (usually with a resigned or fatalistic look about them), fine features, and a lean, brown frame accustomed to heat and hardship.

European Asians live in a region known as South-West Asia. It extends from Afghanistan in the east to Turkey in the west. The people include Turks, Arabs, Iranians and Jews. Much of the land is desert, which makes it almost useless for farming, except near the coasts, but beneath those desert sands lies most of the world's crude oil.

The discovery and exploitation of oil in that area has brought sudden riches to a traditionally poor people and has also resulted in a violent upheaval of long-established ways of life. Wandering desert tribesmen, owning little more than their tents and camels, have seen their burning, trackless homelands transformed almost overnight into gigantic oilfields equipped with all the latest machines and inventions.

European Asians tend to have darker skin and hair than other Europeans. The Arabs, in particular, have thin, somewhat hooked noses and thin lips. Their eyes have that faraway look that comes from centuries of gazing over immense distances against the blazing sun.

The Israelis, who have come from all corners of the earth to carve a new home for themselves out of the wilderness of what was formerly Palestine, have reversed the usual process. They have left their mark on the land in a short space of time by taming and harnessing nature to an impressive degree. With their *kibbutzim* (communal agricultural settlements) they have gone far to fulfil the Biblical prophecy that '. . . the desert shall rejoice and blossom as the rose'.

Top left: Japanese girls wearing the kimono, Japan's traditional dress. Today kimonos are seldom worn outside the home.

Above: A Chinese trader in traditional costume. Again, the Chinese rarely wear such ornate clothes today.

Right: Two women and a child from Bali, one of the most beautiful of Indonesia's many islands.

People of North America

North America is the third largest of the continents. It stretches from the Arctic to the tropics and includes the United States, Canada, Mexico and Greenland. It also includes the countries of Central America and the islands of the West Indies.

The continent forms less than a fifth of the Earth's land area and holds less than a tenth of the world's people, but because of its abundance of natural resources and the skills of its workers it produces about half the world's manufactured goods. The northern four-fifths of North America are occupied by the United States and Canada, and it is these two countries which provide the greater part of the continent's farm and industrial products.

The peoples of North America are descended from four principal sources. The original inhabitants were the American Indians. Historians believe that they arrived in North America from Asia many thousands of years ago. At that time Alaska was linked to Asia by a natural land bridge that would have

A native Alaskan—an Eskimo child—admires a fur coat. The fleshy faces of Eskimos help to keep them warm in the very cold winters of their homelands.

made the crossing easy. After Christopher Columbus had re-discovered America at the end of the 1400s Spanish, English, Dutch and French invaders and colonists gradually killed or imprisoned most of the Indians they found there. Today in the United States and Canada their few descendants may be found carving souvenirs for tourists in special areas called reservations, or in menial jobs in the cities.

The Indians in Mexico and some of the Central American republics were never completely subdued. Today they make up a large part of the population in those countries.

Marriages between Indians and their Spanish conquerors produced *mestizos,* the name for offspring of this kind of mixed ancestry, and *mestizos* are also plentiful today. Both Indians and *mestizos* remain among the poorest of the peoples of Central America.

The Eskimos made up a second source of North Americans. They also came originally from Asia but are believed to have arrived in their new home only some 2,000 years ago. They settled in Greenland, northern Canada and Alaska.

The harsh climate in which Eskimos live forced them to become clever hunters. They survived by killing polar bears, whales, walruses, seals, caribou, musk oxen and other animals, but the Eskimos' way of life has changed in recent years. Many work in the small towns that fringe the Arctic and in the newly-developing oilfields.

Although their ancestors came from the same part of the world, Eskimos are not American Indians. They have similar straight black hair, high cheekbones, dark brown eyes and wide faces, but their skin is lighter than that of the Indians, they have shorter arms and legs, and smaller hands and feet. Their short, stocky bodies and fleshy faces help to keep them warm in the ice and snow.

The African Negro is the third ancestor of the North American peoples. Negroes were taken as slaves from Africa to North America from the early 1500s to the early 1800s. Nearly 10 out of every 100 people in the United States today are Negroes or have mixed white and black blood. Many of the islands of the West Indies also have large Negro populations and some, such as Haiti, are almost entirely black.

American Negroes have dark skin and brown eyes, and their hair tends to be woolly or very curly. Their skin is well adapted to the hot sun and, as a result, they are usually happiest and most often found in the warmer and sunnier parts of the continent where their athletic physiques and easy-going natures are seen at their best in an outdoor existence.

Fourth and most widespread among the ancestors of today's North American peoples were the Europeans. Almost half of all Canadians are of British or Irish descent, and about a third have French ancestry. Immigrants from all over the world have poured into the United States, many as refugees.

Each individual nationality—Poles, Dutch, Chinese and so on—has tended to form its own nucleus in various parts of the country, and contributed its particular talents to the good of the whole. Mexico and the Central American countries were colonised almost exclusively by Spaniards. Their descendants reflect their ancestry in their language, sports (such as bullfighting), dress and religion.

An American Indian in his traditional robes. Indians of many tribes gather at Gallup, New Mexico every August for an Intertribal Ceremony, wearing their finest clothes and performing centuries-old dances.

People of South America

South America is the fourth largest continent, and is almost as big in area as Europe and Australia put together. It is a land of extremes, ranging from the almost unbearable moist heat of the Amazon River Valley to the ice-bound, rocky ground of Tierra del Fuego; and from the rain-soaked forests of Brazil to Chile's Atacama Desert, parts of which have seen no rain for centuries. It is not surprising, therefore, that the lands and peoples of this vast continent are as varied as the climate.

The peoples of South America are increasing in numbers faster than those of any other continent. Most of them are descended from Europeans, American Indians, or Negroes, and many have mixed blood. Most of the people live on or near the coasts because the interior of the continent is difficult to reach, even today, and some parts of it are still unexplored.

About all that the peoples of South America have in common is their religion (more than 90 per cent. of them are Roman Catholics), and their language (almost always either Spanish or Portuguese).

American Indians were the first people of South America. Then white settlers came from Europe, mainly from Spain and Portugal, the Portuguese settling in what is now Brazil. Today Brazilians make up about half the population of the whole of South America.

Many of the white colonists, both Spanish and Portuguese, were drawn to the rolling grassy uplands and treeless plains that cover much of the continent. This type of country is found in Colombia and Venezuela in the basin of the River Orinoco; in parts of Paraguay and northern Argentina where it changes into sparse scrubland with the name of Gran Chaco; in southern Brazil and Uruguay; and in the vast plains of central Argentina which are known as the pampas.

The pampas make up the richest and most productive farmland in South America. To tend the cattle on these never-ending ranges there developed a fiercely independent, hard riding, hard fighting cowboy—the colourful gaucho of Argentina and Uruguay. The gaucho of old was usually a *mestizo* (person of mixed Indian and Spanish ancestry). He spent most of his time in the saddle and was a superb horseman. He ate unsparingly of beef, almost his only food, and washed this down with constant draughts of *maté* (Paraguayan tea) and *aguardiente* (sugar brandy). Today's gaucho is less of a romantic freebooter and more of a hired servant.

Other settlers forsook the land for the sea, and their descendants can be seen in the muscular, tanned fishermen of the north-eastern Brazilian coast, whose frail craft penetrate far into the South Atlantic waters.

A flood of immigrants from Europe arrived in the 1900s. Most settled in the cities, where they found themselves quite happily pursuing trades and professions they had been trained to do in Europe. In places such as Buenos Aires, Rio, São Paulo and Caracas they discovered standards of luxury and convenience that rivalled anything to be found in Europe or North America.

The darker peoples living in South America—Negroes and those of mixed white and black blood—are found mainly in the northern parts of the continent. They are well represented in Brazil and especially in what once were the Guianas: French Guiana, Surinam and Guyana. Many are forest workers or prospectors for gold and diamonds. Others are more conventional mineworkers. Their great strength lies in their being able to tolerate, like the American Indians, hot and humid conditions that would quickly kill or incapacitate a white person.

When the Spaniards and Portuguese first

Opposite: A Quechua Indian of Peru playing a quena, an ancient type of flute made of clay.

Below: A street scene in La Paz, the principal city of Bolivia. In the background is the snow-capped peak of Mount Illimani. La Paz is the highest large city in the world, more than 3,600 metres above sea-level.

arrived in South America in the 1500s they were opposed by large numbers of American Indian tribes. Some of these were peaceable; others were warlike and headhunters. Almost all the Indians were killed in battle or died of diseases introduced by their conquerors. Today the remnants of a few diehard tribes still hide in the depths of the Brazilian rain forests. Bolivia has a large number of American Indians who work mostly in the silver and tin mines, and whose white masters treat them little better than slaves.

Paraguay is unique because the Guaraní

Above left: A large wall-painting done by children in the streets of La Paz, Bolivia. South Americans like bright colours.

Above right: An Otavalo Indian mother and child. The Otavalos form one of several tribes among the Indians who make up more than half the population of Ecuador.

Indians, who make up a large proportion of the population, have achieved sufficient status to have their language accepted along with Spanish for popular, everyday use. Government decrees and newspapers are printed in both languages.

Only in Chile did the native Indian inhabitants refuse to surrender to the Spanish invader. For 300 years the Araucanians fought stubbornly to retain their independence, and only in the late 1800s were they forced on to reservations in south central Chile. About 200,000 of their descendants live there still.

People of Africa

Africa, the second largest continent, has always been regarded as a land of mystery. Because of this it was named the 'Dark Continent'. Since World War II almost all the African countries that had been ruled by European nations have become independent. As a result there has been a dramatic increase in trade, exploration, and all-round development throughout a continent in which seven out of every ten people are black.

The peoples of Africa can be divided into two main groups: Negroes and Caucasians. The Negroid group can be subdivided into true Negroes, Bushmen and Hottentots, Pygmies and Nilotes. The Caucasians can be subdivided into Arabs and Berbers, Europeans and Asians.

True Negroes have skin that ranges in hue from light brown to brown-black. They are distinguished by their woolly hair, broad, rather flattened noses, and thick lips. They live in some of the hottest parts of Africa, where the dark pigment of their skin and thick texture of their hair have evolved as a natural form of protection from the sun's fierce rays.

Above: A group of African children performing a traditional dance, to the accompaniment of clapping hands from their elders.

Left: Two Moroccan women in their best finery. Even today, many women in the Islamic countries of the north have their faces covered: these two are more modern in behaviour.

Most true Negroes live in East Africa, in south-eastern Africa and Madagascar, and in the grasslands and along the coasts of West Africa.

Bushmen are found mainly in and around the Kalahari Desert, in southern Africa. They are wandering hunters who roam the desert in small bands, eating whatever animals they can find and catch, and supplementing this diet with roots, berries, insects and grubs. As might be expected from their harsh, frugal life in the desert, Bushmen are small and wizened, with yellowish brown skin. They have flat faces with high cheekbones, and crinkly hair.

Hottentots resemble Bushmen but are slightly taller. Most of them live in Namibia (South West Africa), where many have married white settlers.

Pygmies are the shortest of all Negroes. Very few exceed 1.4 metres in height. Most pygmies have reddish-brown skin, flat noses, and brown, tightly curled hair on their round heads. Their legs are short and their arms are long. They live mainly in the tropical forests of the Zaïre River basin, where they kill birds and other animals with spears and bows and arrows. The women gather herbs, roots, and fruits for food.

Experts believe that the pygmies' unusual bodies have evolved over thousands of years to enable them to make the best use of the hot, moist forests which are their home. A larger, heavier man would find it difficult to move

Left: A Zulu warrior in full war dress, carrying a shield and an assegai (an iron-tipped spear). Back in the early 1800s the Zulus fought several fierce wars.

Right: A street scene in Cape Town, one of the main cities of South Africa. The children selling bananas are 'Coloured'—that is, of mixed ancestry.

through the dense undergrowth. For this reason, too, the pygmies travel almost naked.

In contrast to the pygmies, the Nilotes are extremely tall Negroes. They range from about 1.78 metres to 2.13 metres in height. There are about 10,000,000 of them altogether, and they live mainly in Kenya, Uganda and Sudan. Slender and graceful, with fine features, most Nilotes are cattle herdsmen. This is certainly true of the Masai tribe, who herd their animals on the open plains in the south of Kenya. Their long legs

Bantu tribespeople in the dry, dusty grasslands of Tanzania.

and prodigious leaping prowess may possibly have evolved from a life on the plains where speed spelled the difference between killing and being killed.

African Negroes live mainly south of the Equator. Arabs and Berbers, who are Caucasians (white), live mainly north of the Equator. Nomad Arabs, called Bedouin, roam the Sahara with their camels, sheep, and goats, ever seeking fresh pastures and water.

Berbers are the earliest known inhabitants of the Mediterranean Sea coast of Africa. They are mostly herdsmen and farmers living in fortified villages. The Tuareg are the Berber equivalent of the Bedouin, living a similar life in the Sahara. Today many Arabs and Berbers have intermarried.

Arabs and Berbers, especially those who live in or near the desert, are lean and suntanned. Their hair is generally curly and their lips and noses are thin. Their lean, hardened bodies are well adapted to resisting a regular round of heat, thirst and sandstorms.

In South Africa a distinct racial group called Cape Coloureds has emerged. These are peoples of mixed European and Negroid blood. In eastern Africa about 1,000,000 Asians, mostly Indians, have made their home. Bringing with them their business skills, they have largely transformed the trading sectors of their communities and brought a welcome prosperity to previously underdeveloped regions of the continent.

127

People of Australia

Australia looks like an island but is really a continent. It is the smallest inhabited continent in the world. It is also the only continent consisting of a single country. It is more than 30 times as large as Great Britain, yet because it has only a little more than 13 million people, almost three-quarters of it is empty.

Australia is unusual in other ways, too. It has a number of animals that are found nowhere else in the world (see pages 44-45). Its extreme contrasts are typified by lush, tropical rain forests in the north-east, burning deserts in the centre, and nearly 2,000 kilometres of mountains whose highest peaks are snow-capped more than half the year.

The first explorers and settlers came from Britain. They met in Australia a strange, gentle, dark-skinned race of Stone Age peoples—the Aborigines. The Aborigines wore no clothes, grew no crops, and had no permanent homes. They wandered over the country living on grubs, insects and small game, which they attacked with throwing-sticks, called boomerangs, and flint-tipped spears.

The descendants of those Aborigines are still there, but out of the 300,000 or so who lived in Australia in the 1700s, only about 47,000 pure-blooded individuals remain, although there are about 93,000 persons of mixed blood. Most Aborigines are to be found on reservations in dry inland regions, many of them living the primitive lives of their forefathers. A few have drifted to the cities, where they hold down menial jobs.

Experts believe that the Aborigines first arrived in Australia some 30,000 years ago, probably from the islands of South East Asia. Thousands of years of outdoor life under the most primitive conditions have endowed the Aborigine with certain strengths and characteristics. Among these are an uncanny tracking ability, superhuman physical endurance and an unshakable fatalism.

Among the original white settlers in Australia were a colony of convicts deported in earlier days from Britain. These men had to be tough merely to survive the sea journey. Those who were eventually freed there found the rugged outdoor existence to their liking and prospered. Subsequent immigrants, mostly from Europe, have been hardy, adventurous types.

Most of the Australian cities grew up hugging the coastline, where the climate is usually kinder and the land easier to subdue. Even there the accent is still on the pioneering spirit, with most families opting for hard work and hard play.

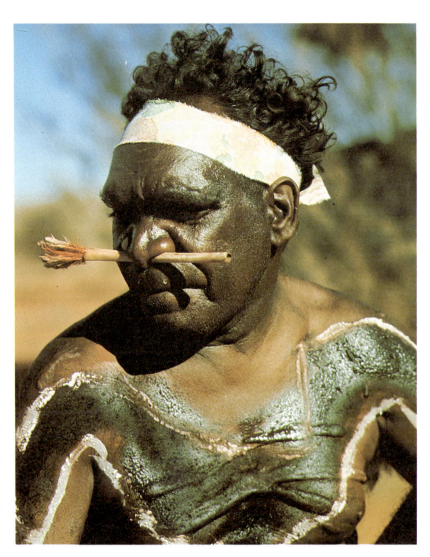

Above: An Aborigine dancer in full finery. Some of the Aborigines still keep to their traditional ways, but many now live and work among the white people of Australia and follow their ways of life and clothing.

Right: A sheep sale at Dubbo, in New South Wales. Australia is the world's leading producer of sheep, and has almost 14 times as many sheep as people.

People of Oceania

Oceania is the name given to some 30,000 islands scattered across the Pacific Ocean. For the purposes of this article, New Zealand is included in Oceania, although many geographers prefer to group it with Australia under the heading of Australasia.

The peoples of Oceania fall into three main groups: Polynesians ('people of many islands'), Melanesians ('people of black islands') and Micronesians ('people of small islands'). Only the Polynesians have enough characteristics in common to qualify as a distinct group.

Generally speaking, Polynesians live on islands lying to the east of the International Date Line, the imaginary line which marks where one date ends and the next begins. They have the lightest skin of all three groups, straight to wavy hair, high cheekbones and brown eyes, and are tall and well-built.

Like the rest of the Pacific Islanders, the Polynesians are thought to have arrived in their present homes many hundreds of years ago from the South-East Asian mainland. They sailed from island to island in frail craft, much like the boats they use today. Their

This group of Maoris from New Zealand are dressed in traditional costume and are performing at a Maori concert party.

descendants, dependent on the sea for much of their food, display the same fearless qualities and superb seamanship.

Today's Polynesians share the same looks, customs and language. Their fine physique and contented outlook may have evolved from an environment (enjoyed by the Maoris of New Zealand, Tongans, Western Samoans and Hawaiians) that was challenging without being harsh, and where food was readily obtainable when sought.

All that Melanesians have in common with each other is that they are shorter and darker than the peoples of the other two groups. Their hair is black and woolly and many look like African Negroes. Some, called Negritos, living in heavily forested areas where height would be a hindrance, are so short that they are classed as pygmies. Melanesians are found in New Guinea, the New Hebrides, Fiji, the Solomon Islands and New Caledonia.

Micronesians are of medium height, have copper-coloured skins and generally straight black hair. Most tribes are unrelated to each other. They are found on the Caroline, Mariana, Marshall and Gilbert Islands.

Languages

Language is the principal way we express our ideas and feelings. We use language to communicate with other people, whether we speak it or write it.

Nobody knows for certain how language began, but one theory is that our first ancestors started to copy the sounds of nature which they heard around them, and these sounds eventually became linked with their source. Each community may have heard and interpreted the same sound somewhat differently, and this would account for the many different languages that are used in the world today. For there are nearly 3,000 spoken languages now, and that total does not include dialects, which are varieties of the same language used by a particular class of people or in a particular district.

The three things that all languages have in common are (a) they are made up of certain sounds; (b) they have rules for changing these sounds into words; (c) the words are arranged according to some kind of grammar.

When writing was invented, about 5,500 years ago, it became possible for the first time to make a record of language. The first written language (used by the ancient Sumerians) was picture-writing. This told a story or gave a message. The alphabet developed when people began to let their picture letters stand for the sounds that they used in their speech, rather than just for the object pictured. The ancient Phoenicians are believed to have started this development. It eventually became standardised in today's alphabets, in which letters represent speech sounds.

Language experts group today's various languages into families of related languages. According to this system, the Indo-European family is by far the largest, its languages being spoken by about half the world's population. The languages in this group are all similar in sound patterns, vocabulary and grammar. They are spoken by Russians, most other Europeans, peoples of the Americas, Australia, New Zealand and northern India. They vary from Catalan to Welsh, and from Urdu to Icelandic.

Within the Indo-European family are eight living branches: Germanic, Romance, Balto-Slavic, Indo-Iranian, Greek, Celtic, Albanian and Armenian. The language you are reading this in—English—belongs to the Germanic group which also includes German, Dutch-Flemish, Swedish, Norwegian, Danish and Icelandic. English is spoken by about 400 million people around the world. Irish, Welsh, Scots Gaelic and Breton belong to a different branch, the Celtic. Romance languages

Opposite, above: An 18th century artist's idea of the Tower of Babel, which according to the Bible story was the origin of the world's different languages.

Above: Children listening to a story on tape at a Melbourne school.

Opposite: United Nations interpreter Theodore Fagan at work. The UN provides translations of all speeches into Chinese, English, French, Russian and Spanish.

include French, Spanish, Portuguese, Italian and Romanian.

There are also many other language families apart from the Indo-European one. About 700,000,000 people speak languages in the Sino-Tibetan family. This group includes Chinese, Tibetan, Thai and Burmese. Arabic, Hebrew, Amharic (spoken in Ethiopia) and the Berber tongues (spoken in North Africa) make up the Semitic-Hamitic-Kushitic family, spoken by about 150,000,000 people.

More than 100,000,000 people speak one or other of the related languages of Hungarian, Finnish, Turkish, Mongol, Manchu and the tongues of Asian Russia, grouped together in the Uralic and Altaic families.

Among smaller families are Basque, which is unrelated to any other known language and is spoken only by the peoples who live in the Pyrenees Mountains, on the borders of France and Spain; and Caucasian, which includes languages spoken in the Caucasus.

A story in the Old Testament of the Bible describes a time when 'the whole earth had one language and few words'. Thousands of years later, the world has come to see the need of a universal language with many words, so that each may speak plainly to his neighbour without the dangers and inconvenience of translation.

More than 600 such 'manufactured' languages have been invented and proposed, but only two have been accepted in any measure. These are Esperanto and Interlingua. The more popular of the two, Esperanto, is spoken by about 8,000,000 people, whereas Interlingua was intended mainly for scientific reports. There is a large section of public opinion that objects to artificial languages on the grounds that they cannot reflect the culture of a people. For them, English is a very real alternative because it is already a compulsory second language in the schools of most of the developed countries of the world.

People and Places Inquiry Desk

Which is the world's greatest river?
The Nile, the mighty river of north-eastern Africa, is the longest single river in the world. It flows 6,700 kilometres from central Africa through the Sudan and Egypt into the Mediterranean. The Amazon River in South America—the world's second longest river—is actually the biggest river. Its length is 6,275 kilometres, but it carries much more water than the Nile. With its tributaries or side branches the Amazon drains an area of nearly 7,700,000 square kilometres. The Mississippi and Missouri rivers in the United States are often considered as one river and have a combined length of 8,150 kilometres. However, the Missouri does not join the Mississippi until about halfway down its length. The longest continuous flow is only 5,100 kilometres. Singly the Missouri is 4,368 kilometres long and the Mississippi is 3,782 kilometres long.

? ? ?

What is the Common Market?
The Common Market is the popular name given to the European Economic Community (EEC). It was founded in 1957 by the Treaty of Rome. At that time the member countries were Belgium, France, Germany, Italy, Luxembourg and the Netherlands. In 1973 Britain, Ireland and Denmark also joined.

The aim of the Common Market was to create a European customs union by removing all trade barriers and restrictions so that members could trade freely with each other. At the same time there was to be a common tariff (tax) system on all goods entering the Market from the rest of the world. By the late 1960s this had been achieved. In time the Common Market hopes to build a United Europe in which goods, people, services and money can all move about without any restriction.

? ? ?

Why does Venice have canals instead of roads?
Venice is built on about 120 tiny mud islands just off the coast of northern Italy. During the AD 400s some refugees from the Roman Empire

Gondolas moored in the canals of Venice, whose buildings loom through the morning mist.

sought safety on these islands from invasions by barbarians (warlike tribesmen from the north). The refugees formed a community, building their houses on wooden piles driven into the mud. Over the years the community developed and by the 800s had become the city of Venice. By the 1300s Venice was the strongest sea-power in Europe.

Today Venice is connected to the Italian mainland by a long road and rail bridge. There are no roads in the old part of Venice, only canals between the many islands. The only way to travel in Venice is on a long, slim boat called a gondola, by motor-boat, or on foot along paths and over bridges.

? ? ?

What is meant by the Third World?
The term Third World refers to the underdeveloped or developing countries of Africa, Asia and Latin America. The phrase was introduced in the 1940s. The world was divided into three. The First World referred to countries such as the United States or Britain. These are countries which are technologically advanced and have so-called free economies. The term Second World refers to countries such as the Soviet Union which have economies that are planned by the State. The term Third World was used to describe the poorer countries of the world such as Chile whose economies were less advanced than the others.

What is the meaning of the term 'population explosion'?

The word 'explosion' refers to the rapid way in which the population of the world is increasing today. From the time that man first walked on Earth until comparatively recently the world's population grew fairly slowly. It was about 500,000,000 in 1650. By 1900 there were about 1,600,000,000 people in the world. Since 1900 the world's population has almost trebled, and because of improvements in medicine and living standards it is growing faster than ever before. Today there are about 4,300,000,000 people. As the world's population is growing by 2 per cent each year it is probable that in 30 years' time it will be more than 8,500,000,000.

Scientists and statesmen are seriously worried about problems such as overcrowding, pollution and the shortage of food and natural resources that such a huge population could cause, and seek ways of slowing down the increase.

Why does nobody live in Antarctica?

Because it is too cold. Antarctica is the fifth largest continent in the world and lies in the area surrounding the South Pole. With its icy winds and massive icebergs, it is the coldest place on Earth. Its winter temperatures can and do drop well below –56°C (–70°F). Hardly any plant or animal life can exist. One species of wingless insect, a few minute organisms and some mosses, lichens and algae are the only permanent inhabitants of this desolate land.

What does the United Nations do?

The United Nations has many activities. Its main task is to try to keep peace in the world. Where there are two unfriendly nations the United Nations can step in to arrange talks. If these fail the United Nations can ask its members not to trade with the countries. Finally, the United Nations can send troops to a trouble-spot. The United Nations has no standing army of its own but can call upon the troops of member countries.

Since the United Nations was formed in 1945 it has intervened in various major conflicts, including the Korean War of 1950-53.

The United Nations is also concerned with starvation and poverty in the world. It has agencies such as the Food and Agricultural Organisation (FAO) and the World Health Organisation (WHO) which deal mainly with the poorer nations, helping them to raise their living standards and to fight disease.

Why are so many cities built on the banks of rivers?

There are many reasons for this. Before the days of railways and good roads, rivers were a very important means of transport, and some, such as the River Danube in Europe, still are. The capital cities of three countries are on its banks—Vienna (Austria), Budapest (Hungary) and Belgrade (Yugoslavia). In more modern times settlers in North America founded cities on the banks of rivers, such as St. Louis on the Mississippi and Washington, D.C., on the Potomac.

Rivers were also valuable for defence. A city on one side of a large stream could not easily be attacked by an enemy on the other side. A river can also be a source of fresh water. London, for example, draws a great deal of its water from the Thames.

The Queen Maud Mountains rise through the ice-cap covering Antarctica, which in places is more than 4,000 metres thick.

Highlights of History: 1

BC

30,000 Appearance of modern man —*Homo sapiens*

12,000 Last Ice Age ends

8000 Agriculture develops in western Europe

6000 Beginning of New Stone Age

3100 Kingdoms of Upper and Lower Egypt united

3000 Bronze Age culture in Mesopotamia; Sumerian civilisation develops along Euphrates River

2700 First pyramid built in Egypt

2500-1500 Indus civilisation, India

2500-1400 Minoan civilisation, Crete

2360 Akkad dynasty founded

2000 Bronze Age, northern Europe

1760 Shang dynasty, China

1750 Babylonian empire founded

1400 Iron Age, India and Turkey

1250 Israelites, led by Moses, leave Egypt

1232 Israelites arrive in Canaan

1100-609 Assyrian empire

814 Phoenicians found Carthage

776 First recorded Greek Olympics

660 Byzantium (Istanbul) founded

563 Birth of the Buddha

559 Persian empire founded

551 Birth of Confucius

509 Roman republic founded

499-478 Ionian Wars; Greeks revolt against Persian rule

479-404 Delian League; Greek states ally under Sparta to oppose Persia

460-429 'Golden age' of Athens

460-404 Peloponnesian Wars: Athens defeats Sparta

336-323 Alexander III 'the Great',

The Battle of Naseby, shown here in a print of the time, was a turning point in the English Civil War of 1642-1646.

The Renaissance

The Renaissance, from a French word meaning 'rebirth', was a period of enormous advances in learning, technology, philosophy and all forms of art. The Renaissance developed in Italy from the 1300s and reached its height during the 1400s and 1500s. It spread throughout Europe and marked the end of medieval thought and values.

---

king of Macedonia, conquers Turkey, Phoenicia, Egypt, Persia and Mesopotamia; invades India

321-184 Maurya dynasty, northern India

264-146 Punic Wars between Rome and Carthage; Carthage destroyed

221-202 Ch'in dynasty unifies China

65-63 Romans invade Syria, conquer Palestine and annex Judaea

58-51 Julius Caesar conquers Gaul and invades Britain

45-44 Caesar becomes dictator of Rome and is assassinated

31 Rome conquers Egypt

5 Birth of Jesus Christ

AD

30 Crucifixion of Jesus Christ

43 Roman conquest of Britain

101 Roman Empire at greatest extent

135 Dispersal of Jews from Israel

220-270 Goths invade eastern Europe

227-641 Sassanid dynasty, Persia

325 First Council of Christian Church, Nicaea, Turkey

360 Picts and Scots invade Britain; Huns invade Europe

395 Roman Empire divided into East and West; capitals Constantinople (Byzantium) and Rome

407 Roman occupation of Britain ends

410 Visigoths sack Rome

476 End of Roman Empire in the West

481-511 Clovis I founds Frankish empire, forerunner of modern France

527-565 Byzantine (Eastern Roman) Empire reaches height

570 Birth of Muhammad, founder of Islam

584 Anglo-Saxon kingdom of Mercia founded in England

618-907 T'ang dynasty, China

634-649 Muslim Arabs conquer Egypt, Syria, Mesopotamia and Persia

661-750 Omayyad dynasty, Islam

680 First Bulgarian empire founded

700 Arabs conquer Tunis, North Africa

712 Arabs found Sind, Pakistan

757 Papal States founded, Italy

800 Charlemagne, King of the Franks, crowned Holy Roman Emperor

856-875 Vikings invade Britain

878 England divided between Alfred, King of Wessex, and the Danes

896 Magyar tribes conquer Hungary

900 Hausa kingdom founded, Nigeria

920-1050 Empire of Ghana at its height

930 Europe's oldest parliament—the Althing—assembled in Iceland

962 Holy Roman Empire re-founded by Otto I, King of Germany

987-1441 New Mayan Empire

1054 Eastern (Byzantine) Church independent of Rome

Major Wars

Peloponnesian War	431-404BC
Punic Wars	264-146BC
Hundred Years War	1337-1453
Wars of the Roses	1455-1485
Thirty Years War	1618-1648
English Civil War	1642-1646
War of the Spanish Succession	1701-1713
Seven Years' War	1756-1763
American War of Independence	1775-1783
Napoleonic Wars	1792-1815
War of 1812	1812-1814
Crimean War	1853-1856
American Civil War	1861-1865
Franco-Prussian War	1870-1871
World War I	1914-1918
Spanish Civil War	1936-1939
World War II	1939-1945
Vietnam War	1957-1975

1055-1071 Seljuk Turks conquer most of Asia Minor

1066 Normans invade England; William I, first Norman King of England

1075-1122 Struggle between popes and German emperors over appointment of bishops

1096 First Crusade to free Palestine from Turks

1147 Yoritomo, first shogun (military governor) of Japan

1167-1227 Mongols under Genghis

An unknown Dutch artist painted this picture of the Great Fire of London in 1666. The fire destroyed more than 13,000 houses, churches and other buildings.

Khan conquer northern China, Siberia and Persia (Iran)

c1200 Aztecs settle in Mexico

1200-1450 Hanseatic League—trading association of German cities

1200-1533 Inca civilisation, Peru

1210 Franciscan order of monks founded

1215 Magna Carta guarantees feudal rights of English barons
Dominican order founded

1240 Mali empire founded, West Africa

1241 Mongol kingdom set up in Russia

1260 Kublai Khan, Mongol ruler, founds Yüan dynasty, China

1271-1295 Venetian traveller Marco Polo visits China

1273 Rudolf I, Holy Roman Emperor, first Habsburg emperor

1291 Saracens capture Acre from Christians; Crusades end

1295 First true English Parliament held by Edward I

1314 Scotland independent after Battle of Bannockburn

1334-1351 Black death—plague in Europe

1337-1453 Hundred Years' War between England and France; England loses all French possessions except Calais

1368-1644 Ming dynasty, China

1378-1417 Great Schism divides Roman Catholic Church

1381 Peasants' revolt, England

1397 Union of Kalmar; Denmark, Norway, Sweden united

1440 Invention of printing

1453 Ottoman Turks capture Constantinople; rise of Ottoman Empire

1455-1485 Wars of the Roses; civil wars in England between houses of York and Lancaster

1478 Spanish Inquisition begins

1479 Castile and Aragón united; Spain becomes one country

1485 Wars of the Roses end; Henry VII, first Tudor King of England

1488 Portuguese navigator Bartolomeu Dias sails round Cape of Good Hope

1492 Christopher Columbus reaches West Indies

1517 Protestant Reformation begins

1519-1521 Portuguese sailors make first voyage around the world

1519-1521 Hernán Cortés of Spain conquers Mexico

1526-1858 Mughal Empire in India

1534 Act of Supremacy; Henry VIII breaks with the Pope

1540 Society of Jesus (Jesuits) is founded

1545 Counter-Reformation begins

1547 Ivan IV, the Terrible, crowned as first Tsar of Russia

1558 French capture Calais; last English possession in France

1558-1603 Elizabeth I, Queen of England

1562-1598 Wars in France between Roman Catholics and Huguenots

1588 Spanish Armada defeated by English

1607 Colony of Virginia founded

1609 Netherlands gain independence from Spain

1618-1648 Thirty Years' War, Europe

1620 Pilgrim Fathers sail to North America

1642 New Zealand discovered by Dutch navigator Abel Tasman

1642-1646 Civil War in England between Charles I and Parliament

1649 Charles I executed

1652 Dutch found Capetown, South Africa

1652-1674 Anglo-Dutch Wars

1653 Oliver Cromwell becomes Lord Protector of England, Scotland and Ireland

1665-1666 Great Plague and Fire of London

1667-1669 War of Devolution between France and Spain

The Industrial Revolution

The Industrial Revolution started in Britain during the mid-1700s. During the next hundred years Britain changed from a farming country to an industrial one. Machines came into use, factories developed and industrial towns sprang up. By the 1830s industrialisation was spreading to most of western Europe and the United States.

1688-1689 Glorious Revolution in England; Parliament offers throne to William III and Mary II

1700-1721 Great Northern War between Sweden and other states

1701-1714 War of the Spanish Succession

1701 Union of Scotland and England

1721 Robert Walpole becomes Britain's first Prime Minister

1733 John Kay invents flying shuttle: Industrial Revolution begins

1733-1735 War of the Polish Succession

1739 War of Jenkins's Ear between Britain and Spain

1740-1748 War of Austrian Succession

1756-1763 Seven Years' War, Europe

1759 British take Quebec, Canada from the French

The Reformation

The Reformation is the name given to the religious movement that swept Western Europe in the 1500s. It began as a protest against abuses in the Christian Church and was strongly influenced by the Renaissance revival of learning. The Reformation resulted in the establishment of Protestantism and other dissenting religions.

Highlights of History: 2

1770 James Cook discovers New South Wales, Australia

1775-1783 American War of Independence

1783 Montgolfier brothers build first successful hot-air balloon

1787 United States adopts constitution

1787-1792 Russia and Turkey at war

1788 First convicts arrive in New South Wales from Britain

1789 French Revolution begins

1789-1797 George Washington, first President of the United States

1791 Quebec is divided into two self-governing territories, Upper and Lower Canada

1808-1814 Peninsular War in Spain and Portugal

1809-1825 Wars of Independence in Latin America

1812 Napoleon invades Russia

1812-1814 Britain and United States at war

1814 Allied forces invade France; Napoleon abdicates

1815 The Hundred Days; Napoleon returns to power but is defeated at Battle of Waterloo

1819 *Zollverein* (customs union) established, Germany
Singapore founded by the British

1820 Liberal revolutions in Spain, Portugal and Italy

The mud and devastation of World War I: Australian troops passing along a duckboard through the remains of Château Wood, near Ypres in Belgium, in 1917.

1791 Negro slaves in Haiti revolt against the French

1792 France proclaimed a republic

1792-1802 French Revolutionary Wars

1793 Louis XVI, King of France, executed

1795-1796 Scotsman Mungo Park explores Gambia and Niger rivers, Africa

1801 Great Britain and Ireland united as the United Kingdom
Islamic Sokoto kingdom founded, West Africa

1803 Louisiana Purchase; France sells Louisiana to United States

1804 Bonaparte crowned Napoleon I, Emperor of France

1805-1815 Napoleonic Wars

1806 Holy Roman Empire dissolved

1806-1812 Turkey at war with Russia

1807 Slave trade abolished in the British Empire

1820-1821 Egypt conquers Sudan

1821-1830 Greek war of independence from Turkey

1822 Liberia, West Africa, founded for freed American slaves

1823 Monroe Doctrine warns against European interference in Americas

1824-1827 British Ashanti wars, West Africa

1824-1826 Anglo-Burmese War

1825 First passenger railway opens, England
Trade unions legalised in Britain

1830 July Revolution, Paris

1830-1831 Belgium independent

1832 British Reform Act extends vote to middle class

1834-1839 Carlist war in Spain

1836 Texas wins independence from Mexico

1836-1837 Great Trek; Boers settle in Transvaal, South Africa

1837-1901 Victoria, Queen of Britain

1838-1848 Chartist movement in Britain seeks political reform.

1838 Boers defeat Zulus in South Africa

1839-1842 Opium War between China and Britain

1840 New Zealand becomes a British Crown Colony

1840 Upper and Lower Canada united

1842 Hong Kong ceded to Britain

1845-1848 Anglo-Sikh Wars, India

1845-1847 Potato famine in Ireland: a million people die

1846-1848 United States and Mexico at war

1848 Year of revolutions in Europe

1850-1864 T'ai P'ing rebellion, China

1851 Great Exhibition in London

1852-1870 Second Empire, France

1853-1856 David Livingstone crosses Africa

1854-1856 Crimean War; Britain, France and Turkey against Russia

1857-1858 Indian Mutiny; British government takes over India from East India Company

1861 Unification of Italy

1861-1865 American Civil War

1862 Slavery abolished in U.S.A.

1867 Dominion of Canada established

1869-1878 Cuban revolt against Spain

1869 Suez Canal opened

1870-1871 Franco-Prussian War

1871 Unification of Germany

1876 'Custer's Last Stand'—Battle of the Little Big Horn

1880-1881 First Boer War

1882 British troops occupy Egypt

1883 The Mahdi, a fanatic, leads revolt in Sudan

1884-1895 European powers partition Africa and create colonies there

1887-1896 Italy and Ethiopia at war; Ethiopia regains independence

1896-1898 Anglo-Egyptian forces reconquer Sudan

1898 Spain and United States at war; Cuba gains independence

1899 First Hague Peace Conference

1899-1902 Second Boer War

1900 Boxer Rebellion, China

1901 Australia becomes an independent dominion

World War I

World War I began in 1914 when Austria attacked Serbia in revenge for the assassination of an Austrian prince by a Serbian student. Germany, Bulgaria and Turkey joined Austria, and Russia, Britain, France and Italy backed Serbia. Millions died in fighting by land, sea and air. The war ended in 1918 when Germany and its allies surrendered.

1903 Suffragette movement begins in Britain

1903-1906 Norwegian explorer Roald Amundsen sails through North-West Passage

1904-1905 Russo-Japanese War

1905 Revolution in Russia; limited reforms granted

1907 New Zealand becomes an independent dominion

1909 American explorer Robert Peary reaches the North Pole

1910 South Africa becomes an independent dominion
Japan annexes Korea

1911 Italo-Turkish war; Italy occupies Tripoli, Libya

1911 Norwegian explorer Roald Amundsen reaches the South Pole

1911-1912 Revolution in China; Manchu dynasty overthrown

1912-1913 Balkan wars; Balkans partitioned

1914 Irish Home Rule Act passed

1914-1918 World War I

1916 Easter Rising, Ireland

1917 Russian Revolution; Bolsheviks seize power
Balfour declaration promises Jews a national home in Palestine

1918 Germany becomes a republic

1919 Fascist party founded in Italy

1920 Palestine comes under British administration
League of Nations formed

1922 Irish Free State set up
Turkey becomes a republic

1924 First Labour Government in Britain

1926 General strike in Britain

1927 Civil war in China between Communists and Nationalists
American Charles Lindbergh makes solo transatlantic flight

1929 United States' stock market collapses; world-wide economic depression begins

1931 Japan seizes Manchuria from China

1933-1934 National Socialists gain control in Germany; Adolf Hitler becomes Führer

1935 Italian forces invade Ethiopia

1935 Persecution of Jews begins in Germany

1936-1939 Spanish Civil War

1938 The Anschluss: Germany and Austria united

1939-1945 World War II

1946 United Nations founded

1946-1954 Civil war in Indochina

1947 Independent states of India (Hindu) and Pakistan (Muslim) created

1947 Palestine divided into Arab and Jewish states

World War II

World War II began in 1939 when Germany and Russia invaded Poland. Britain and France declared war on Germany. By 1941 Germany controlled most of Europe, and then attacked Russia. The same year Japan attacked Britain and the United States, and fighting became world wide. Germany was defeated in May 1945, and Japan surrendered a few months later after atomic bombs devastated two of its cities.

The German dictator Adolf Hitler rides in triumph into Vienna in 1938.

1948 State of Israel established

1948-1949 Israel and Arab League at war

1949 Germany divided into East and West

1949 North Atlantic Treaty Organisation (NATO) formed

1949 Communist régime in China

1950-1953 Korean War

1951 Chinese occupy Tibet

1952 Mau Mau terrorists fight British in Kenya

1953 Egypt becomes a republic

1954-1975 Vietnam War

1955 Soviet bloc forms Warsaw Pact

1956 Uprising in Hungary suppressed by Soviet Union
Egypt nationalises Suez Canal; Anglo-French intervention fails

1957 European Economic Community (Common Market) formed
Soviet Union launches first space satellite, Sputnik 1

1959 Communist régime in Cuba

1960-1975 Most African countries achieve independence

1961 Soviet astronaut Yuri Gagarin becomes first man in space

1962 Soviet-US confrontation over Cuba

1964 United States' involvement in Vietnam War increases

1965 Rhodesia declares unilateral independence from Britain

1967 Six-Day War; Israel defeats the Arab states

1967-70 Civil war in Nigeria

1968 Terrorism begins in Northern Ireland
Soviet troops invade Czechoslovakia

1969 American astronauts land on the Moon

1971 East Pakistan independent as Bangladesh after civil war

1973 Arab-Israeli war

1974 Watergate scandal in USA

1977 President Sadat of Egypt starts peace talks with Israel

History Inquiry Desk

Which were the earliest civilisations?
Civilisation means 'living in cities', and the earliest civilisations began once man stopped being a wandering hunter and began to settle and farm land. By 6000 BC farming had begun in Western Asia. The first civilisations slowly developed in fertile river valleys in Western Asia, North Africa, India and China.

Probably the earliest civilisation of all developed between the Tigris and Euphrates rivers in an area known as Mesopotamia. There in about 5000 BC

tribes who invaded Europe in the AD 200s and overran the Roman Empire in the mid-400s. The term was taken from a Greek word meaning stammering and described the sound of the tribes' foreign languages.

There were many barbarian tribes, most of them Germanic or Scandinavian. The Huns were originally from Asia, and they arrived in Europe about 370. They were strongest in the mid-400s under their great leader Attila. The Goths were Germanic people. In the 200s they split into two

was organised as a pyramid with the king at the top. Beneath the king were his great barons and bishops. Below them were the lesser nobles and land-owners. At the bottom was the mass of serfs and peasants.

The basis of this structure was land, which was granted by the most powerful to the less powerful in return for service. (The word 'feudal' comes from the word 'fee' used to mean payment, in service or money, for land.) In this way the king gave his land to his nobles, and they fought for him when needed. Further down in the social scale the freeman paid rent for his land, while the serf was virtu-

Early civilisation: This impression of the seal of King Darius of Persia shows the king in his chariot hunting lions in the 500s BC.

some people called Sumerians settled. They founded the world's first city-states and developed the first known system of writing. From about 2300 BC they declined and gave way to the great Assyrian and Babylonian empires.

Other early civilisations included the Indus River Valley civilisation in what is now Pakistan and the cities of ancient Egypt.

? ? ?
Who were the barbarians?
The word barbarians was used by the Romans to describe the northern

groups—the Ostrogoths (East Goths) and Visigoths (West Goths). They settled between the Don and Dneister rivers and in the Ukraine. Another important tribe was that of the Vandals, who sacked Rome and brought the Roman Empire to an end in 456.

? ? ?
What was the feudal system?
The feudal system was the way in which society was organised and governed in Western Europe during the Middle Ages. The system was first developed by the Franks of France.

Under the feudal system society

ally a slave working his lord's land; in return his lord was expected to protect him. Although feudalism began to fade away from the 1300s, it continued in Germany until the 1800s and in Russia until 1917.

? ? ?
What was the Long March?
The Long March was a journey made by the Chinese Communists led by Mao Tse-tung in 1934-1935. From 1917 there had been civil war in China between the Communists and Nationalists. In 1931 the Communists established a republic in Kiangsi Pro-

vince. It was attacked by Nationalist forces under Chiang Kai-shek, and in 1934 the Communists were forced to travel more than 9,600 kilometres to Shensi Province on the Yellow River. Some 90,000 marchers began the journey but only half survived. Despite a brief truce fighting continued. In 1949 Chiang and the Nationalists were forced to flee to the island of Taiwan (Formosa), leaving the Communists in control of mainland China.

？ ？ ？
What was the Holy Roman Empire?
The French writer Voltaire once said that the Holy Roman Empire was

Right: A poster showing Mao Tse-tung, the Chinese leader from 1949 until he died in 1976. Below: Captain James Cook.

neither holy, nor Roman, nor an empire! The name was given to a federation of many central European states. Although its borders changed many times, the Holy Roman Empire contained what are now Austria, Germany, Switzerland and northern Italy.

It had its origins under Charlemagne, king of the Franks, who in 800 tried to revive the Western Roman Empire, which had ended in 456. In 936 the German king Otto I revived the idea again, and became the first of a line of emperors which lasted until 1806. The title 'Holy Roman Emperor' was first used in 1157. From the end of the 1200s the emperor was elected from the house

(family) of Habsburg, the rulers of Austria. The Empire was brought to an end in 1806.

？ ？ ？
Who discovered Australia?
The Dutch sailors Willem Jansz and Abel Tasman first sighted Australia during the 1600s. They named the area New Holland. It was the British explorer Captain James Cook who was the first person to chart Australia and most of the Pacific islands. In 1768 he set out on an expedition to Tahiti. His voyage lasted until 1771, during which time he sailed round New Zealand and landed on the eastern coast of Australia.

？ ？ ？
What was the Great Trek?
The Great Trek is the name given to a historic journey made by the Boers (Dutch settlers) in South Africa in 1835-1837. By 1806 Britain had taken over Cape Colony in southern Africa from the Dutch. The Dutch resented British rule and in 1835 some 10,000 Boers set off northwards. They settled in Natal and founded the Orange Free State and the Transvaal. By 1887 Britain had annexed all three Boer republics. War broke out between the Dutch and British in the Boer Wars of 1880-1881 and 1899-1902. In 1910 the entire country was united as the Union of South Africa.

Science and Technology

You probably take it for granted that your home has electricity and water. People rely to a large extent on supermarkets with their conveniently packed foods. Washing machines, wallpapers, cars, computers, television, radios, cameras and many other goods play an important part in our lives.

Yet only a few hundred years ago people lived entirely without these things. They have become possible by the incredible advances in technology—the use of machines to do work. Technology has, in turn, been made possible by the ideas of scientists.

The uses of electricity provide an excellent example of how inventions have followed scientific discoveries. In the 1800s scientists learned how to use and control electricity. This led to the invention and production of batteries and electric motors. When electricity became available in the home it was quickly adapted for use in lighting. The electrical inventions that followed included the tele-

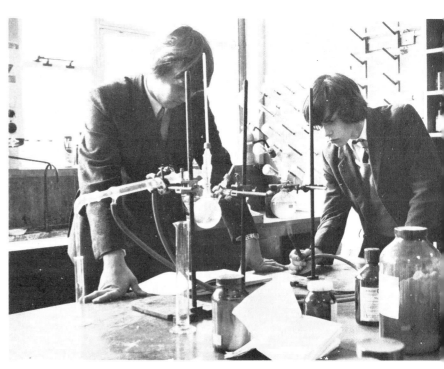

Above: The experiment these boys are carrying out in their school laboratory will help them to reach a greater understanding of science.

Science and technology in action: refining crude oil and (foreground) a filling station selling some of the finished product—petrol.

phone, heating appliances, radio, television, calculators and all the other electrical goods we have today. The science of electronics, a branch of electricity, is still progressing at a great pace.

The word 'science' literally means 'knowledge', but it is more than just a gathering of facts. A scientist observes things and from his observations decides on a probable reason for their cause. This is called a theory or hypothesis. The scientist then tests the theory by carrying out as many experiments as possible. If the theory passes all the tests then it is regarded as a scientific law. However, many theories are disproved by experiments.

Many of the early scientists each tackled a whole range of subjects such as mathematics, astronomy, optics and medicine. Since then the amount of scientific knowledge has increased tremendously. As a result scientists

140

now tend to concentrate on smaller areas of specialised study, and science is divided into several main branches. Physics is the study of matter and energy; chemistry is the study of how substances are made up and how they react with each other; biology is the study of living things; medicine is the science of healing; geology is the study of the Earth; astronomy is the study of the stars and planets; and mathematics is the study of numbers and their applications.

All these branches of science are inter-linked: no branch can be studied completely on its own. For example, the study of medicine involves much biology, particularly human biology, which in turn requires a knowledge of some chemistry and physics. There are areas of study that involve both biology and geology. Physics and astronomy also overlap, and both require an understanding of mathematics.

The main branches of science are also divided into even more specialised areas of study. For example, biology is broadly divided into zoology (the study of animals) and botany (the study of plants). Biochemistry is an increasingly important aspect of biology, and much of the latest information about animals and plants has come from the study of

Engineering plays a big part in modern technology, as shown in these huge metal spheres and the forest of pipes at an oil refinery.

the chemical processes that go on in the cells which make up their bodies.

Physics is also divided into areas of study. Mechanics is concerned with the effects of forces upon objects; thermodynamics is concerned with heat and its effects; optics is the study of light; acoustics is the study of sound; electricity and magnetism are studied together; and nuclear physics is concerned with the physical properties of atoms.

There are two main branches of chemistry. Inorganic chemistry is concerned with elements and their compounds found in minerals and other non-living things. Organic chemistry is concerned with the compounds of carbon, many of which are found in living organisms.

Using the discoveries of scientists, industry now supplies us with a large number of our needs. Energy from coal and oil is often converted into electricity, but our supplies of coal and oil cannot last for ever and so alternative sources of energy are being sought. Technology has given us goods made of plastics, metals, and materials from living organisms, and we have found ways of using our resources to provide ourselves with various forms of transport. At the same time problems such as waste-disposal have arisen. These too must be solved by technology.

Oil and Natural Gas

The petrol and diesel fuels used in aeroplanes, cars, trucks and trains are the main products obtained from oil. A surprising number of other materials, such as plastics, insecticides and animal foods, are also made from oil and natural gas.

Oil and natural gas are found in a number of places below the surface of the ground. They were formed millions of years ago from the remains of once-living organisms. When these organisms died their bodies collected in shallow pockets at the bottom of the seas. They were quickly covered by sediments, such as sand and mud, which hardened to form rocks. Meanwhile the dead organisms were decaying into the chemicals of oil.

Over the years the oil seeped through the porous rocks until it came to an impermeable (non-porous) layer of rock. There it formed a pocket of oil we now call an oil field. Frequently gas was formed at the same time as oil, and pockets of natural gas are often found above oil. In some cases only gas is found beneath the ground.

Geologists know where to look for oil because there are certain rock formations in which oil traps are frequently found. Upward

Above: Geologists carrying out a seismic survey at work in the forest at Umutu, in Nigeria. The country has one of the richest reserves of oil in Africa south of the Sahara.

Prospecting for oil in Alaska, a geologist takes notes on the shores of a glacier-fed lake. In the background is Mount St. Elias, one of Alaska's highest peaks.

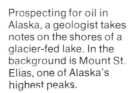

bulges in the rock, called anticlines, make excellent oil traps. Sometimes rocks have been broken and have slid against each other, forming a fault. In a fault where an impermeable rock has slid against a porous rock, an oil trap is often found next to the impermeable rock. Salt domes—underground masses of salt—are also impermeable, and oil is often found where a salt dome has pushed the overlying rocks upwards.

Geologists find these rock formations by a technique called seismic surveying. They set off small explosions on the surface of the ground, and these explosions send shock waves down through the rocks. The shock waves are reflected in different ways by the various layers of rock underneath. By recording the reflected waves geologists can build up a picture of the underground rocks. From such pictures, called seismographs, the geologists can say whether the rocks below are likely to contain oil or not.

Having selected a likely area, the only certain way to discover if oil is present is to drill for it. Shallow oil wells are drilled by percussion drilling, in which the drill-head hammers through the rock. However, most oil wells are drilled by rotary drilling, in which a revolving drill is gradually pushed through the rock. As the drill progresses downwards lengths of drill pipe are added by men on the

drilling platform. The hole is lined with a metal casing, which is forced down with the drill pipe. If oil is discovered the drill pipe is removed and the hole is capped with cement. When the oil is needed a pipeline is connected and the cement cap is pierced. The well then becomes a producer.

The crude oil that comes out of the ground is piped to a refinery. There the chemicals contained in the oil are separated from each other, and some of them are changed into other chemicals.

The first stage in refining is called fractional distillation. The crude oil is heated in a furnace and then passed to a tall tower, called a fractionating tower. In this there are a number of interconnecting levels. The gases and the most volatile liquids (those that evaporate most easily) can reach the top levels, from where they are removed. This very light fraction includes petrol (gasoline) and naphtha. Heavier fractions are removed lower down the tower, and the residue—the heaviest

Oil rigs like this are needed for prospecting for petroleum and natural gas under the sea, and for tapping the supplies when they are found. Work on an oil rig is hard and often dangerous.

fraction of all—is removed from the bottom of the fractionating tower.

The amount of petrol produced by fractional distillation is not enough to meet the enormous demand. Therefore, some of the chemicals in the lower fractions are broken down into the chemicals of petrol. This process is called catalytic cracking, and it involves splitting up large molecules into smaller, more volatile molecules.

Fractional distillation and catalytic cracking also produce chemicals suitable for use in making plastics, laboratory chemicals (such as acetic acid and toluene), insecticides (such as DDT), antifreeze (ethylene glycol), detergents and animal foods.

The chemicals of natural gas can also be separated. Some are used as fuels in the form of bottled gas, such as butane and propane. Others are used in industry to make methyl alcohol and synthetic rubbers. Unlike crude oil, however, natural gas can be used directly as a fuel in the home.

Generating Power

The people of the Middle Ages would have regarded electricity as magic. Even the early scientists of the 1600s and 1700s would have described the idea of sending pictures through space as pure science fiction. Yet today science fiction has become science fact. We can send television pictures from the Moon to the Earth, and we rely considerably on electricity in our everyday lives.

Electrical phenomena were known as long ago as the 1500s, but it was not until the 1800s that scientists discovered how to produce electricity and began to understand its nature. Today we know that electricity is caused by free electrons (see pages 164-165) moving through a conductor, such as a metal wire.

There are numerous ways of causing this movement of electrons. The earliest sources of electricity were cells and batteries, first made in the early 1800s. Since then other methods of producing electricity have been discovered, including generators and solar cells.

Electricity was first thought of as an invisible fluid. As a result its movement became known as a current. A source of electricity was described as having a force that made electricity move—an electromotive force (emf). Today electric current is measured in amperes, and the emf is measured in volts. Hence the emf is more usually called the voltage.

The strength of an electric current depends on the voltage and also on the resistance to the flow of current, which is measured in ohms. A good conductor, such as a metal, has a lower resistance than a poor conductor. At the same time resistance decreases with an increase in the cross-sectional area of the conductor. Thus a thick wire, such as a power cable, has a lower resistance than a thin wire.

Batteries and many generators produce an electric current that flows in one direction only. This is direct current (DC). The electricity supply to your home is in the form of alternating current (AC). The current builds up to a maximum flow in one direction, falls to zero, and then builds up to a maximum flow in the other direction. This change in direction takes place very rapidly. In domestic supplies 50 or 60 complete cycles occur every second.

There are good reasons for having alternating current instead of direct current. First, it can be generated more efficiently. Second, it can be transformed from one voltage into another, which is important in distributing electricity. Also, alternating current can be easily converted into direct current if needed for some direct-current motors, which work more efficiently. Converting direct current to alternating current is more difficult.

The electricity supplied by power stations is produced by generators. These machines work on the principle of electromagnetism. In 1821 the English scientist Michael Faraday discovered that when a wire moves through a magnetic field an electric current is induced in it. A modern generator contains several coils of wire, which are made to rotate in a strong magnetic field.

The energy needed to rotate the coils may come from one of several sources. In countries with a plentiful water supply electricity is produced in hydro-electric power stations. The water is dammed and then released through tunnels containing water turbines. The force of the water turns the turbines, which are linked to the generators. However, most generators are powered by steam, which is produced by burning coal, oil, peat, or natural gas, depending on which is most convenient. Nuclear power stations use the heat produced in their reactors to make steam.

Electricity is distributed in what is called the grid system. This consists of power stations and substations, all linked by power cables, so that electricity can be distributed to

Opposite, top: This huge rope-driven alternator is one of a group installed in a London power station back in 1889. Sections were replaced periodically from 1895 onwards. Opposite, below: The forty 12.5 cm ropes which drove each of the alternators. The drive came from giant flywheels powered by two steam engines.

Reservoirs of water trapped behind huge dams provide the energy to generate power in hydro-electric stations like this one at Otago, New Zealand.

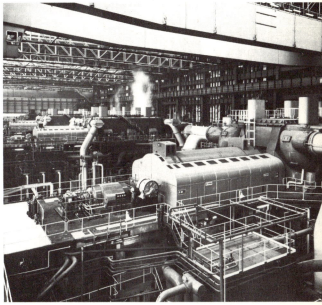

wherever it is needed. Electricity is produced from a power station at about 25,000 volts, but to distribute it at that voltage would result in a great power loss in the cables. Therefore, the voltage is raised by transformers to between 132,000 and 400,000 volts, depending on the distance it has to travel. At these higher voltages the loss in the cables is very small. In North America, where the distances are very great, the voltage may be as high as 750,000 volts to minimise loss.

However, such voltages are too high for normal use, and so transformers at local substations reduce the voltage to levels suitable for use in towns, industry and the home—between 110 and 240 volts.

The turbine hall in a modern power station. It contains four 500 megawatt turbo-generator units. The power to drive them is provided by coal.

Coal from Ancient Swamps

A cutting and loading machine at work on a long coal face, deep underground in a mine. This mine has coal-face lighting for greater safety.

About 300 million years ago there were vast, swampy forests of plants in many parts of the world. The underground seams from which we now extract our coal are the remains of these plants. When they died they decayed under special conditions and gradually became compressed. Frequently we find fossil impressions in the coal, and these give us some idea of the plants and animals that lived during this time.

Coal is sometimes described as a mineral, but this is not accurate. It consists mostly of carbon compounds, derived from the ancient plants, together with small amounts of sulphur, phosphorus and rock minerals. When the plants died they dropped into the stagnant water of the swamp and did not decay completely. A slimy material was formed, which became compressed into peat.

This process can still be seen going on in boggy areas where peat is formed, for example in Ireland. There sphagnum moss dies and decays in the stagnant water of the bog, while new moss plants grow above. Gradually a great thickness of dead moss material forms.

Coal was formed when the peat became

even more compressed. Over many millions of years the sea covered and uncovered the land several times. Each time a layer of sediment was left behind, and these layers later hardened into sedimentary rocks. The pressure of the overlying rock compressed the peat into a light form of coal called lignite.

In places where the overlying rocks were thicker, the coal became even more compressed and formed bituminous coal. In some places even more drastic changes occurred in the rocks. They became buckled and folded, and the extreme pressures that this caused formed anthracite, the hardest coal of all.

Coal is extracted from the ground by mining. Where it lies near the surface it can be removed by opencast mining. The layers of soil and rock above the coal are removed to expose it. Sometimes an opencast mine reaches a point where the layers of rock are too thick to be removed economically. In such cases augers (large drills) are used to remove the coal by boring holes into the coal seams. When an opencast mine has been worked out the rock and soil can be replaced. Within a few years the land can be used again for other purposes.

Coal measures (seams) that lie deep in the ground have to be reached by underground mining. Often a vertical shaft is sunk and the coal is mined by tunnelling outwards from the shaft along the coal seams. Miners and equipment are lowered to the coal seams by a hoist in the main shaft. A second shaft is also sunk so that air can be pumped through the mine.

Sometimes it is not necessary to sink a vertical shaft. In a slope mine the coal face is reached by a sloping tunnel. This makes it easier to transport coal to the surface. A drift mine is used when the coal is buried in the middle of a hill. A horizontal tunnel is bored into the side of the hill.

Coal can be burnt, and since its discovery it has been much used as a fuel, particularly in the days of the steam engine. However, coal has many other uses. By a process called carbonisation it can be broken down into four main constituents. The coal is heated in large ovens from which air is excluded. When it reaches a temperature of about 1200°C, it is converted to coke, coal-tar, ammonia and a mixture of gases we call coal gas.

The coke itself can be used as a fuel, and is particularly used in smelting iron in blast furnaces. Coal gas was used in the home, and still is in places where supplies of natural gas are unobtainable. The ammonia can be used to make fertilisers.

Coal-tar contains a large number of useful chemicals. They can be used to make many products, such as plastics, dyes, drugs, disinfectants and weedkillers.

Right: From prehistoric forest to modern chemicals—the story of coal over millions of years.

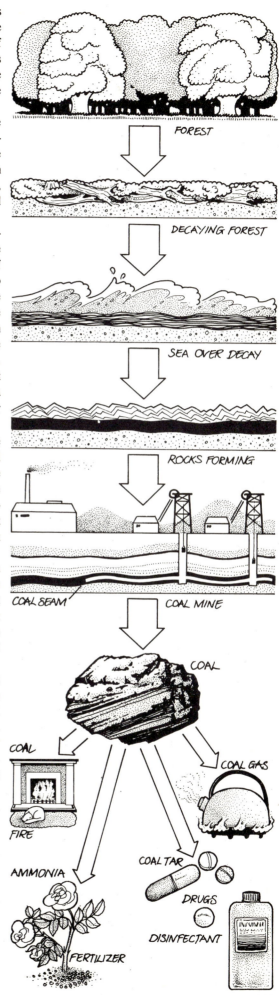

FOREST

DECAYING FOREST

SEA OVER DECAY

ROCKS FORMING

COAL SEAM   COAL MINE

COAL

COAL FIRE

COAL GAS

AMMONIA

FERTILIZER

COAL TAR

DRUGS

DISINFECTANT

Alternative Energy

Most of the world's energy comes, at the moment, from the fossil fuels—oil, natural gas and coal. Oil and natural gas supplies may run out in the next 50 years and even coal cannot last forever. The main alternative energy supply is nuclear power, but scientists are looking at other ways of supplying energy.

The heat and light of the Sun, wind, water and the heat from inside the Earth are all possible sources of energy. Many people are in favour of creating an alternative technology to nuclear power, but the main problem is to harness these sources to provide sufficient energy at an economical price.

The Sun gives out a vast amount of energy, most of which is lost in space. Some falls on the Earth, where it is essential for the existence of life. Much of this energy is radiated back into space, but scientists have devised two ways of trapping a little of the Sun's energy for our own use.

Heat from the Sun can be trapped in solar panels. A domestic solar panel consists of a thin, black container filled with water and angled to catch as much sunlight as possible. The Sun's heat is absorbed by the water, which is pumped away through pipes. These pipes pass through a water tank in which the heat is transferred to the domestic water supply in a system known as heat exchange.

Windmills like this were used to grind corn in the years before the Industrial Revolution. The framework of the sails was covered with cloth to catch the wind. The top of the windmill could be turned round so that the sails always pointed into the wind for maximum power.

An array of solar panels set on a sloping roof. An array of this kind can provide enough power to heat domestic water supplies even in a climate where there is not a great deal of hot sunlight.

Having lost heat by warming up the domestic supply, the water in the pipes returns to the solar panels for reheating.

Solar panels are in use all over the world. Obviously they are most useful in countries where the Sun shines nearly all the day. In cooler countries such as Britain they are used to heat swimming pools and greenhouses and to aid the existing hot water supply.

Solar cells are a means of converting the Sun's light into electricity. A solar cell consists of a substance such as selenium sandwiched between two layers of glass. The Sun's light causes the selenium to generate an electric current. The main problem is that a single cell only produces a very small electric current. To produce large currents enormous numbers of solar cells are needed. Thus at the moment they are used in photographic light meters, where only small electric currents are required, and in space satellites, where other power sources are unsuitable.

The inner layers of the Earth are hot, and sometimes this heat comes to the surface in the form of hot lava, hot springs and geysers. In New Zealand, the United States and Iceland electricity is generated using this kind of

geothermal power. Engineers harness the steam that comes up through cracks in the ground and use it to drive steam turbines.

Cold water is yet another possible source of energy. Hydro-electric power stations on rivers are already much in use, but the tides and waves of the sea also contain energy. Tidal power is being used in the River Rance estuary in France. There is a barrage across the estuary and as the tide rises and falls the water runs through pipes that contain water turbines.

One metre of an average ocean wave contains enough energy to power a hundred homes. As yet, no one has developed a satisfactory method of harnessing this energy, but various experiments are being carried out. One of the latest ideas is a wave raft, which consists of a series of hollow pontoons hinged together so that they rock up and down. Hydraulic rams link the pontoons at the hinges, and as the rams move in and out they operate generators.

Windmills have been used for centuries, particularly for grinding flour and pumping water. Traditionally they are large, clumsy structures with four huge sails to catch the

Power from inside the Earth is being harnessed at the Wairakei Geothermal Borefield in New Zealand's North Island. Boreholes trap steam from deep in the Earth's crust and lead it through pipes to drive electrical generators.

wind. Modern windmills, however, are very different and are used to generate electricity. They have long, thin vanes and are able to make use of even the lightest winds.

The main problem with electricity-producing windmills is to ensure that the electricity output does not rise and fall with the wind speed. The largest windmill in the world is in Denmark, at Tvind. It is computer controlled and the vanes automatically adjust their angle so that the windmill keeps revolving at the same speed whatever the speed of the wind. A smaller windmill at Aldborough in England works in a different way. Instead of being directly linked to a generator it lifts a load of 15 tonnes over a height of 1.2 metres. The generators are powered by hydraulic pressure as the load descends.

An ingenious wind experiment is being carried out in New Zealand. There a scientist has succeeded in operating a pump by linking it by ropes to the tops of four trees. As they sway in the wind, the tops of the trees do not all move together. As a result, there is a constant up and down movement where the four ropes meet. Perhaps we may soon see tree-powered electricity.

Building a Home

We build houses to provide ourselves with shelter, so that we can be warm and comfortable. The first houses were built about 25,000 years ago from large stones and the bones of mammoths. Later housing materials included mud, grass, wood, clay, stone and mud-bricks. Modern houses, however, contain a large number of materials, many of which were unknown 50 years ago.

The first stage in building a modern house is laying the foundations. These are essential if the house is to remain standing for any length of time. First, deep trenches are dug where the walls are to be. The bottom of each trench is filled with concrete, called the footings. Above the footings bricks or stones are laid. Stone used to be a popular building material but it is now very expensive. Most modern houses are built with bricks.

Bricks are laid in layers, or courses, on top of the concrete footings. They are laid alternately to give the walls strength, and they are held together with mortar—a mixture of cement and sand. When the bricks reach 15 centimetres above ground level the trenches are filled in, and so the foundations are complete.

At this stage the damp-proof course is laid. This is to prevent dampness in the ground from rising up the walls. It consists of a layer of bituminous felt laid on top of the foundations, 15 centimetres above the level of the ground. A damp-proof membrane is also laid inside the house. If the floor is hollow the membrane is laid on low brick walls, called sleeper walls. These support the wooden joists below the floor-boards. In a solid concrete floor the membrane is buried in the concrete, which protects it from damage.

The walls of the houses are built on top of the foundations. External walls are built as double walls—two walls separated by a cavity. This keeps damp away from the inside wall and helps to insulate the house. Internal walls are only one brick thick. In many new houses composition blocks, such as breeze blocks (made of cement, gravel and ash), are used instead of bricks for the internal walls. They are also used, above the foundations, for the inner layers of cavity walls.

As the walls of the house rise, the door and window frames are built in. If a second storey is being built wooden joists are set into the walls and laid across the house. More wooden joists form the ceiling of the second storey, and finally the sloping rafters of the roof are added. These are covered with bituminous felt, and slates or clay tiles complete the roof.

If there is a chimney a sheet of zinc or lead, called a flashing, is fixed to it and laid over the surrounding tiles. This is to prevent rainwater from running down the sides of the chimney into the loft. Below the eaves, or edges, of the roof, gutters are fixed to the wall to catch rainwater from the roof.

Some houses are built without bricks. Pre-fabricated houses are assembled from factory-made panels and can be put together very quickly. If a frame house is being built a wooden sill is attached to the top of the foundations, and a frame of wooden supports is joined to the sill. The wall sheathing—made of wood, plaster-board or fibreboard—is then nailed to the supports. Planks of wood, called clap-board, are fixed to the outside of the house. The doors and windows are built into the wooden frame.

The gas and water supplies to the house are attached to the nearest gas and water mains. These are underground and the supply pipes to the house are buried in the foundations. Inside the house one water-pipe draws off cold water for the kitchen sink and another, called the rising main, feeds water into the tank in the loft. The gas supply is fed to appliances through a meter. Gas supplies are not always available for houses away from towns.

The drains from the house may lead to the local main drainage system. In more remote areas they empty into either a cess pit, which has to be emptied a few times each year, or a septic tank, in which bacteria break down the sewage and the water soaks harmlessly away into the ground.

The electricity supply is connected to the nearest main cable, sometimes overhead, sometimes underground. In the house the wires are connected first to a meter and a fuse box. A fuse breaks if too many appliances are used at the same time or if an appliance becomes faulty and causes a short circuit. This prevents the wiring itself from overheating and causing a fire. Sometimes circuit-breakers are used instead of fuses.

When the house is built and the services have been connected, the inside can be finished off. Floorboards are laid; plaster-board or fibreboard forms the ceilings, and the walls are plastered with gypsum-plaster. The house is now ready for the remaining fittings, such as the bath, sinks and cupboards, to be installed, and for decorating with wallpaper and paint.

Even after it is finished, a house takes some months to settle down. For one thing, several tonnes of water are used to mix the cement and plaster, and this has to dry out. The timber also takes time to dry completely.

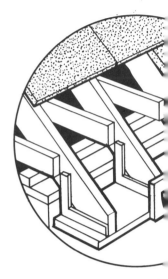

This cutaway picture of a typical modern house shows the details of its construction. Inset below: How the eaves and gutter are fixed to the ends of the rafters.

The house is built with cavity brick walls to keep out the damp. Inset below: Detail of the foundations, showing the damp-proof course just above ground level, and one of the metal ties that link the two parts of the cavity wall.

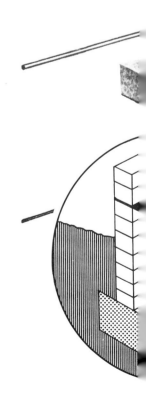

Inset left: Detail of a chimney, showing the metal flashing which covers the gap between chimney and roof.
Inset right: How weatherboarding is nailed on to a wooden frame.

Inset right: Fitting a **reinforced concrete sill and beam** into a brick **wall**. Note the **damp-proof layer**.

Water

Water is the most abundant chemical in the world, which is fortunate because life could not exist without it. It covers 70 per cent. of the Earth's surface, and most of this water is in the oceans. Living things require a great deal of water; the human body consists of 65 per cent. water, a tree is made up of 50 per cent. water, and a non-woody plant contains 75 per cent. water.

Water is an unusual liquid in its behaviour, and its peculiar properties make it extremely useful. For example, within a relatively narrow temperature range (0°-100°C) it can be a solid (ice), a liquid or a gas (water vapour). This is due to the structure of water molecules, which are made up of hydrogen and oxygen atoms.

In each molecule two hydrogen atoms share their electrons with a single oxygen atom (H_2O). However, the atoms are not arranged in a straight line. Instead, the molecule is bent into a shallow V-shape, with the oxygen atom forming the bottom of the V. The shared electrons are not arranged evenly around the molecule, and it is this that gives water its eccentric properties. The electrons tend to accumulate near the oxygen atom, and this causes the oxygen atom to have a slight negative charge, whereas the hydrogen atoms have a slight positive charge. Negative and positive charges attract each other and so water molecules tend to cling together, linked by what are called hydrogen bonds.

The existence of hydrogen bonds thus explains the properties of water. A drop of water on a dry piece of glass tends to form a rounded shape instead of spreading. This is due to surface tension. The surface molecules are being pulled together, preventing the drop of water from collapsing. Another example of surface tension is shown by the fact that a dry pin will float on the surface of undisturbed water.

Water has extraordinarily high melting and boiling points. By rights it should behave like hydrogen sulphide (H_2S), which is a gas at room temperature, but the strong attraction between water molecules tends to prevent them from moving apart to form vapour. However, at a temperature of 100°C the energy of the molecules overcomes the hydrogen bonds sufficiently to allow water to boil.

Other properties result from the positive and negative charges on different parts of the molecules of water. Many chemicals can easily be dissolved in water because they too are made up of charged units. For example, salt (NaCl) molecules are composed of positively-charged sodium atoms (Na^+) and negatively-

Have you ever wondered when you turn on a tap how the supply reaches your home, and how much water is used every day? You consume only a few litres a day in food and drink, but you use probably 25 times as much in washing and flushing drains. Industry uses much more—for example, it takes about 150 tonnes of water to make a tonne of steel.

charged chlorine atoms (Cl⁻). When salt dissolves in water the charged atoms are pulled apart, and in this state they are known as ions.

On the other hand, water is not a good solvent for chemicals that cannot be split up in this way, such as fats and oils. In order to dissolve these substances chemicals called soaps and detergents (see pages 156-157) have to be added to the water.

The water on the Earth's surface is continually being recycled. It evaporates from the sea and condenses in tiny droplets to form clouds. When clouds contain too much water it falls to the ground as rain, snow or hail. Some of this water flows into streams, rivers and lakes, and eventually reaches the sea again. Some water sinks into the ground, from where it can be taken up by plants. Underground there is a nearly constant level of water known as the water table.

We require water for many purposes in the home and in industry. It is generally collected in reservoirs, which may be formed by damming rivers. There the solid material in the

Big cities require huge reservoirs to hold supplies of water for their needs. This reservoir is part of the Murray River, near Albury, New South Wales, and the water is dammed by the Hume Weir. The water is used for irrigation as well as in homes and factories.

The water cycle: Water in the sea and from growing plants is evaporated by the heat of the Sun. It condenses to form clouds, and falls back to Earth as rain. The water soaks into the ground, but much of it flows back into the sea.

water is allowed to settle. Then the water passes through a filter-bed to remove the fine particles. The clear water is sterilised by adding chlorine; if it is lacking any desirable chemicals, such as lime or fluorine, these may be added also. The purified water is stored in enclosed reservoirs or towers until it is needed.

The demand for water is enormous and so engineers are constantly seeking new sources. Some water is obtained by pumping up ground water, but this can lead to a fall in the water table, which may cause settling of the land and any buildings that are on it. In parts of the United States the water table has to be replenished artificially through bore-holes at times when water is plentiful.

Other possible sources of water include the sea and icebergs. Desalinating (removing salt from) seawater is expensive at the moment, but may prove more economical in years to come. It has also been shown that it is feasible to tow icebergs from the Antarctic to dry areas of the world, such as parts of Australia.

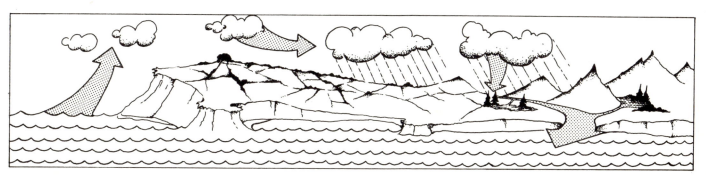

Perils of Pollution

A cement factory at the port of Rufisque in Senegal pours forth smoke and dust over the surrounding land.

Below: At Caracas in Venezuela garbage is disposed of by burning. This avoids one form of pollution, but the fires produce new forms because owing to the way in which the wind swirls among the surrounding mountains, vile-smelling smoke and ash cover the city, particularly at night. Hundreds of scavenger birds await the opportunity to strip carcases.

Water and air are essential for the survival of all forms of life on Earth, including humans. Unfortunately, one of the disadvantages of modern technology is that it has resulted in the pollution of both water and air. Some forms of pollution are under control, but we have by no means solved the problem yet.

Raw sewage is a major source of pollution in a number of countries. If treated correctly sewage can be made harmless, but if it is pumped directly into a river the bacteria that feed on it use up all the available oxygen in the river water. As a result the river becomes lifeless, foul-smelling and a hazard to health.

In some places sewage is pumped into the sea. If winds and tides carry it away from the coast there is no health hazard, but there are places, particularly in the Mediterranean Sea, where bathing is now banned due to the risk of infection from sewage.

Sewage treatment involves allowing the bacteria to break down raw sewage before it is released into a river. The sewage passes through a series of tanks in which the heavier solid matter and the thick sludge are removed. The thick sludge is pumped to a digester tank, where it is digested by bacteria. The digested material can be used as a fertiliser. The more-watery sludge is treated in an activated sludge tank. The sludge is aerated and the oxygen in the air allows millions of bacteria and protozoa to grow. The bacteria feed on the sludge

and the protozoa feed on the bacteria. The digested sludge is again allowed to settle and the remaining water, which is free from harmful bacteria, can be pumped into the river.

Sewage is not the only pollutant of water. Oil spillages from tankers and undersea pipelines can kill large numbers of marine animals and birds. Chemicals from rubbish tips may seep into the ground and eventually find their way into rivers and lakes. Fertilisers that contain nitrates and phosphates may also be washed from the land. They cause accelerated growth of certain water plants, such as algae, which rapidly remove oxygen from the water, leaving it stagnant and unable to support animal life.

Pesticides and industrial waste may kill animals more directly. Waste oils, detergents, poisons such as arsenic and cyanide, and compounds of poisonous metals, such as cadmium, mercury and lead, are all harmful, especially when released in large amounts. Sometimes these poisons become concentrated along a food chain. The small animals that originally take in the poisons are eaten by larger animals, which in turn are also eaten. The animal at the top of the food chain gets the strongest dose of the poison. DDT, a pesticide that is now banned, is known to have been accumulated by a number of animals, including humans, often causing considerable harm. One species threatened in this way is the bald eagle, national bird of the United States. DDT has affected its egg-laying.

Poisoning of humans in Minamata, Japan, resulted from mercury pollution from a petro-

Pollution by accident: oil discharged from the wreck of the supertanker 'Amoco Cadiz' covered the beaches of Brittany in northern France in 1978. Young volunteers from all over France helped to clear the beaches, but thousands of birds and sea-animals died from pollution.

chemical factory. The mercury was accumulated by fish and shellfish in Minamata Bay.

Pollution of the air is no less serious than water pollution. Until 1956 London frequently suffered from smog, which killed millions of people. This occurred when fog—tiny droplets of water—accumulated around smoke particles, making the air thick and unfit to breathe. The 1956 Clean Air Acts, which created smokeless zones, have eliminated smog in Britain, though smog still affects other cities in the world. However, smoke on its own is also a pollutant. It cuts off a large amount of sunshine and forms grime in many cities. Sulphur dioxide, another product of burning fuels, also affects our cities. It reacts with water to form sulphuric acid, and this attacks stonework and corrodes metals.

Another form of air pollution comes from the fumes emitted by vehicle exhausts. These contain carbon monoxide, lead and unburned chemicals of petrol. Carbon monoxide is a lethal gas. If it is breathed in only two parts per thousand it will kill. Lead is also poisonous and it accumulates in the body.

The unburned petrol chemicals are not poisonous, but they react with nitrogen oxides in the air to form what are called photochemical oxidants. These chemicals cling to dust particles in the air and smog is formed. This occurs particularly in Los Angeles, in California, where the geography of the surrounding land prevents sufficient air movement to blow the smog away. In the United States and Japan there are now strict controls on the emissions of petrol engines.

155

Soaps and Detergents

Water by itself can dissolve chemicals such as salt. There are many chemicals that do not dissolve in water and which have to be removed by the action of soap or a detergent.

Soap has been used for many centuries. The Romans were the first to use it in any quantity, but it was possibly invented by the Phoenicians, the Canaanites of the Bible, who lived at the eastern end of the Mediterranean Sea 3,000-4,000 years ago. The invention may have happened when fat from a boiling pot of food spilled into the ashes of the fire beneath. Ashes contain an alkali, and this would have reacted with the fat to form a crude soap.

One of the common fats used in the manufacture of soap is stearate fat. In this substance each molecule (see pages 164-165) consists of three stearic acid units attached to a single glycerine unit. When stearate fat is boiled with an alkali, such as caustic soda (sodium hydroxide), the fat molecule breaks up to give a molecule of glycerol and three molecules of sodium stearate—soap.

A molecule of sodium stearate consists of a long chain of carbon atoms (with hydrogen and oxygen atoms) and a single sodium atom at one end. When soap dissolves in water, it splits up into sodium ions (Na^+) and stearate ions ($C_{17}H_{35}COO^-$). The negative charge is at one end of a stearate ion. As a result, this end is attracted to the positively charged parts of the water molecules (see pages 152-153). The charged end of the stearate ion is thus described as being hydrophilic (water-loving); the other end is hydrophobic (water-hating).

The action of soap depends entirely on the difference between the opposite ends of the stearate ions. When oil is shaken with water, it forms an emulsion of small globules. If the emulsion is left to stand, the globules soon join up and the mixture returns to being a layer of oil floating on water.

However, if soap is added to the mixture the hydrophobic ends of the stearate ions try to escape from the water by burying themselves in the oil globules. The hydrophilic ends remain outside the globules, which thus become surrounded by stearate ions. As a result, they cannot join up again when the emulsion is left to stand.

The action of soap on greasy dirt is exactly the same. Dirt on a fabric is surrounded by stearate ions and lifted away from the fabric.

Soaps have one major disadvantage. Water and dirt often contain calcium or magnesium ions. Indeed, in hard water areas there are

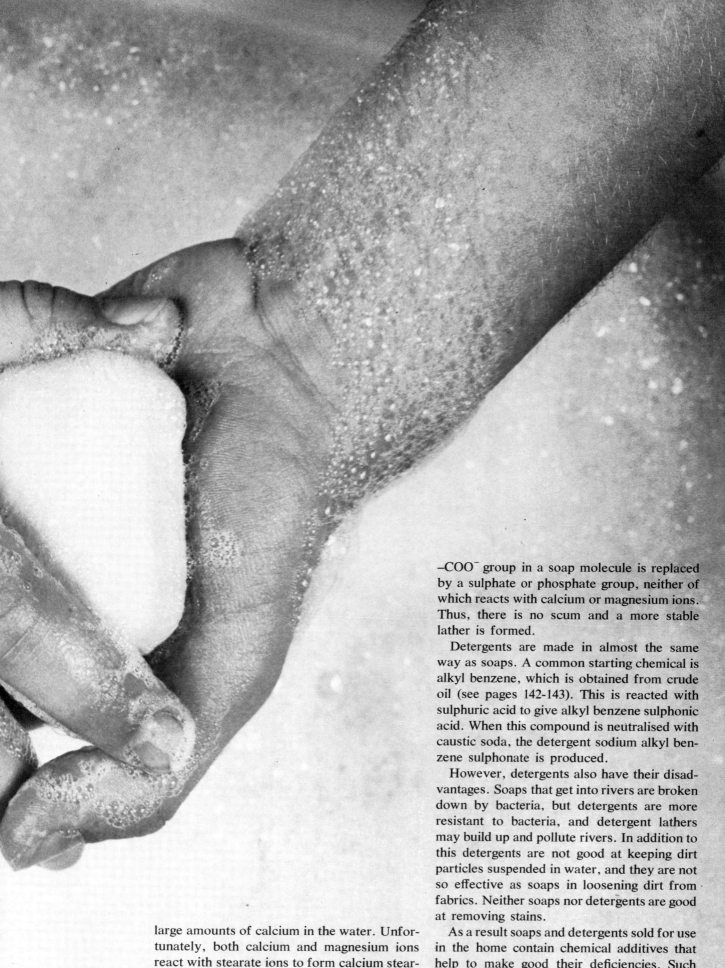

−COO⁻ group in a soap molecule is replaced by a sulphate or phosphate group, neither of which reacts with calcium or magnesium ions. Thus, there is no scum and a more stable lather is formed.

Detergents are made in almost the same way as soaps. A common starting chemical is alkyl benzene, which is obtained from crude oil (see pages 142-143). This is reacted with sulphuric acid to give alkyl benzene sulphonic acid. When this compound is neutralised with caustic soda, the detergent sodium alkyl benzene sulphonate is produced.

However, detergents also have their disadvantages. Soaps that get into rivers are broken down by bacteria, but detergents are more resistant to bacteria, and detergent lathers may build up and pollute rivers. In addition to this detergents are not good at keeping dirt particles suspended in water, and they are not so effective as soaps in loosening dirt from fabrics. Neither soaps nor detergents are good at removing stains.

large amounts of calcium in the water. Unfortunately, both calcium and magnesium ions react with stearate ions to form calcium stearate or magnesium stearate. These substances are both insoluble in water and thus form scum, which can be difficult to rinse out.

Detergents overcome this problem. The

As a result soaps and detergents sold for use in the home contain chemical additives that help to make good their deficiencies. Such chemicals include bleaches, enzymes, water softeners, suds stabilisers, tarnish inhibitors (in dish-washing detergents), whiteners, perfumes and dyes.

Science Inquiry Desk-1

Why don't we just take the salt out of sea-water for drinking?

The main reason is cost. There are several ways of desalinating water (taking the salt out of it) but they all need energy. If you heat salt water the steam produced is salt-free, but this process needs a lot of energy. If you freeze it you can collect the ice and melt it down to get fresh water; this process requires less energy. Other methods involve passing electric currents or high-frequency sound waves through the salt water, but they are all too expensive for commercial use.

? ? ?

Who made the first computer?

The earliest kind of machine for doing sums was the abacus, a frame containing beads strung on wires. It was invented thousands of years ago, probably in China, and is still used. Several mathematicians made calculators in the 1600s and 1700s, but the modern calculator was largely invented by Charles Babbage, an English mathematician, in 1830.

Babbage never quite finished his machine, but many other successful calculators were built in the 1800s and

a motor-car work by friction. They transform the forward motion of the bicycle and car into heat, and so slow the vehicles down.

? ? ?

What is pre-stressed concrete?

Pre-stressed concrete is a kind of reinforced concrete. By itself concrete—which is a mixture of cement, sand and stones—is very strong under pressure—you can pile heavy weights on it—but is lacking in girder strength—that is, when it is supported at both ends but not in the middle. If you reinforce the concrete by casting it around steel rods it is much stronger. Pre-stressed concrete is

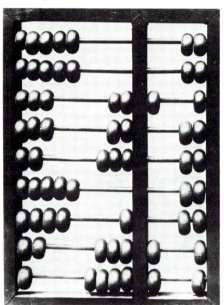

Left: The abacus is one of the oldest computing devices. Right: A modern computer installation.

? ? ?

How do you measure power?

The earliest way of measuring power was devised by the Scottish engineer James Watt, who made the first successful steam-engines. In the 1700s the main sources of power were horses, so Watt coined the term 'horsepower'. He calculated that an average horse walking at $2\frac{1}{2}$ miles an hour (4 kph) could pull a load of 150 lb (68 kg). Multiplying the number of pounds by the distance moved (in feet) gave an amount of work called a foot-pound, and the resulting sum was 550 foot-pounds a second. This, said Watt, was one horsepower.

In the metric system the unit of work is the joule, which equals about 0.74 of a foot-pound. Power is measured in watts, and one watt equals one joule of work a second.

158

early 1900s. The first modern digital computer was built by Professor Howard Aiken at Harvard University in 1944.

? ? ?

Why do people's hands get warm when they rub them briskly together?

The heat is caused by friction, which is the property that stops things sliding about. Without friction you would not be able to walk about—your feet would slip from under you, as they do if you try walking on ice, which has very little friction.

The rubbing of one thing on another is caused by using energy—you use energy to rub your hands together. Energy is never lost, it is merely turned into something else. In the case of friction the energy is turned into heat. The brakes of a bicycle and

made by substituting steel cables for the rods, and stretching them tight, like the strings of a violin, before you pour the concrete round them. The cables compress the concrete end to end for extra girder strength.

? ? ?

What is kinetic energy?

Physicists, people who study matter and energy, divide energy into two basic kinds. Kinetic energy gets its name from a Greek word meaning 'to move', and is energy in motion. A ball flying through the air and any machine in action are examples of kinetic energy.

The other kind of energy is potential energy, and it represents energy that has not gone into action but can do so. If you climb up a slope with a skateboard you have created potential

The fierce heat produced by the Sun is shown in this picture of its corona (ring of flame) during an eclipse.

energy, which becomes kinetic energy as you coast downhill.

? ? ?

How does the Sun produce its heat, and will it ever grow cold?
The Sun is an enormous nuclear reactor. It works the opposite way round to the power-stations at present operating on Earth. We produce nuclear energy by splitting atoms. The Sun's energy comes from nuclear fusion—combining the atoms of hydrogen, the lightest gas, to make helium, which is the next heaviest. In time scientists hope to produce nuclear power in a similar way.

The Sun must eventually grow cold, but its stores of hydrogen are so vast that it can go on giving out heat for thousands of millions of years.

? ? ?

Why is it necessary to build a dam for a hydro-electric power plant?
A hydro-electric power plant works with energy supplied by water moving downhill—in other words, a river. The rushing water, a source of kinetic energy, presses against the vanes of a turbine, turning the shaft of the turbine round and so driving an electric generator.

Many rivers do not fall fast enough to drive a big turbine, though there is enough water in them for this purpose. If you build a dam to hold the water back you level out the river for part of its course, and then let it fall with a whoosh, strong enough to drive your machines.

Most rivers flow much faster in wet weather than they do in dry times, so you would not get an even flow of water all the year round. A dam makes a huge reservoir, or store of water, which is a form of potential energy. So when there is plenty of water in the river you can store it up until the reservoir is full, and let some of the stored water out in dry weather to make up for the shortage. If there is too much at any time you can let it out through spillways which bypass the turbines.

? ? ?

When was petroleum first discovered?
Believe it or not, about 5,000 years ago. The Assyrians and other ancient peoples of the Middle East used bitumen, a tarry form of petroleum, to stick stones together, and as medicine. Alexander the Great burned petroleum as a war weapon to frighten his enemies in the 330s BC.

The Chinese and the Arabs used natural oil for lighting for hundreds of years. The Venetian merchant Marco Polo, who spent many years in the service of the Emperor Kublai Khan of China in the late 1200s, came across what he called 'a fountain of oil' at Baku near the Caspian Sea. The local people, he wrote later, used it in their lamps.

The petroleum industry as we know it today began in the 1850s.

159

Fun with Calculators

A few years ago pocket calculators were expensive and only a few people bought them. Now they are so cheap that a great many people use them—and they are being allowed into the classroom. You can even use them in some examinations, and many teachers claim that they not only make mathematics easier, but help the pupils to understand more about sums.

The reason why calculators are so small and so cheap is a little thing called a silicon chip. Silicon is a semiconductor—that is, it conducts electricity better than an insulator such as glass, and not so well as a metal such as copper. Manufacturers build up complete electrical circuits with layers of silicon and

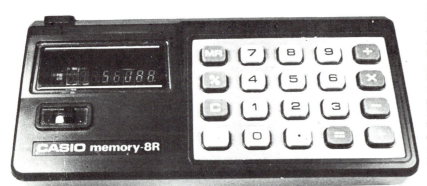

other materials, using photography to make them tiny. A chip a centimetre long can contain the equivalent of many hundred assorted transistors, resistors and capacitors, all of which are used in electronic devices. Every calculator has one of these little chips inside it.

As far as the working of a calculator is concerned, you must remember that in theory a calculator cannot do anything that someone with enough knowledge of mathematics cannot do on paper. The difference is that the calculator does it many times faster.

It is impossible to describe exactly how a calculator works, because different models vary so much in the way they work. It is possible to say something about the various facilities a calculator offers.

All calculators have the digits 0 to 9, arranged in the same pattern, just as the basic keys of a typewriter always follow the same pattern. They also have a decimal point, and the operations add, subtract, multiply and divide. In general, these operations combine together just as you would when doing a sum. For example, to calculate $51.3 \div 2.93$ you put into the machine 51.3, press the divide key, put in 2.93, and press the equals key to get the

result: 17.5085. With more complicated calculations, the order in which you carry out the various stages is determined by the particular model you are using.

Apart from these basic operations, a calculator may work out such things as square roots, powers (such as $2^3 = 2 \times 2 \times 2$), and the trigonometrical functions, sine, cosine and tangent. Many models have a memory, in which a figure can be stored to be used at a later stage.

Most of the practical uses of a calculator are fairly obvious, such as adding up figures, working out costs and so on. More interesting are applications in mathematics and science. For example, suppose you want to work out the square root of a number, and your calculator does not have a square root key. (To remind you, the square root of 9, written $\sqrt{9}$,

Left: A simple calculator. It performs the four basic arithmetical functions, and has in addition a memory key and a percentage key. Below: A more complex calculator, capable of solving many scientific problems. Each key has more than one function. In addition, different programmes can be used to suit the user's needs.

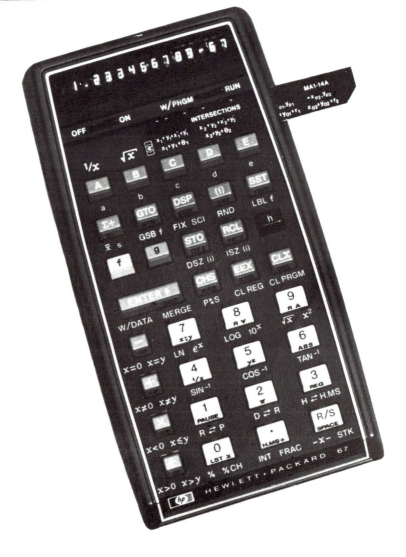

Left: Calculators today are finding a place in the schoolroom. Below: Inside a calculator, showing the printed circuit board that links the keys to the vital silicon chip.

$(2+3) \times 1 - 4 = 1$
$2 \times 3 \times 1 - 4 = 2$
$(2 \times 4 + 1) \div 3 = 3$

How many different numbers can you get?

GETTING CLOSE

Player A uses the calculator first. He uses the numbers 1, 2, 3, 4 once each, together with the operations of the caculator, to get a number, which need not necessarily be a whole number. He should make a note of the process used, for checking purposes.

B then takes the calculator and, using the same numbers, tries to get a result as close to A's as possible, in a given time (say 30 seconds). A scores the difference between his result and B's closest result. A and B then change rôles and repeat the process. The game continues until one player reaches 20 (or whatever total you decide on), and wins.

Example: A goes first:
$[(4 \times 2)^2 + 1] \div 3 = 21.6666$
B then tries to match it:
$4 \times 2 \times 3 - 1 = 23$
A scores $23 - 21.6666 = 1.3334$ for this turn.

is the number which when multiplied by itself gives 9, so $\sqrt{9} = 3$ because $3 \times 3 = 9$.) To find a square root you can use an *iterative* (repetitive) method, in which you start with a rough answer and use it to get a more accurate one.

Suppose you want $\sqrt{7}$. This is very roughly 3, because $3 \times 3 = 9$.

Divide 7 by $3 = 2.3333$.

Add to this figure the original $3 = 5.3333$ and divide by $2 = 2.6667$. This figure is a better estimate of $\sqrt{7}$ than 3 is. (Check: $2.6667 \times 2.6667 = 7.1113$.)

To get a still better estimate, repeat the calculation, using 2.6667 instead of 3:

$7 \div 2.6667 = 2.625$
$2.625 + 2.6667 = 5.2917$
$5.2917 \div 2 = 2.6458$

which is a better estimate of $\sqrt{7}$ than 2.6667 (Check: $2.6458 \times 2.6458 = 7.0003$). You can continue in this way to get better estimates of $\sqrt{7}$ by repeating the process. The limits of accuracy are fixed by how many figures your calculator can show. You can never get an accurate value for $\sqrt{7}$, but since $\sqrt{7}$ cannot be worked out accurately by any method, this does not really matter.

There are many interesting games and puzzles which use calculators. Here are two simple examples:

ONE TO FOUR

A game for two players. Using the digits 1, 2, 3, 4 exactly once each, and the operations of your calculator, try to combine them to make as many whole numbers from 1 upwards as possible. For example:

Using a Microscope

A microscope is an essential piece of equipment for a scientist. The unaided human eyes can see fairly small objects; with a simple magnifying glass, or hand lens, they can see small objects more clearly; but with a microscope they can pick out very fine detail.

Biologists use microscopes to study the cells of plants and animals and the tiny single-celled organisms that cannot be seen with the eyes alone. Chemists use microscopes to examine the fine structure of metals, and geologists use them to study the particles that make up rocks.

The first microscope was made by a Dutch spectacle-maker, Zacharias Janssen, in 1590. It was a compound microscope, with a concave (inward curving) lens next to the eye and a convex (outward curving) lens the other end. Further microscopes were not built until scientists actually needed them for their work.

The Italian biologist Marcello Malpighi (1628-1694) used microscopes to study the blood vessels of frogs and other animals. Anton van Leeuwenhoek (1632-1723), a Dutch biologist, examined anything he could put under the glass, and was the first man to see bacteria. The microscopes these two men used were excellently made, but they only contained a single lens. Robert Hooke (1635-1703), an English physicist, re-invented the compound microscope with two lenses. With this he could examine even finer detail than Malpighi or Leeuwenhoek. Hooke was the first person to see that cork is made up of tiny box-like compartments, which he called cells.

A compound microscope consists of an objective lens near the object and an eye-piece lens, through which you look. The objective lens produces a slightly magnified image of the object. It works in the same way as a hand lens. Light rays from the object are focused on the other side of the lens. The image is called a real image because the light rays that enter your eye are coming directly from the image.

In a compound microscope the eye-piece lens prevents these light rays from reaching the eye by the shortest path. The eye-piece lens changes the direction of the rays, which then appear to be coming from a much larger image. The image you see is thus called an apparent image.

This kind of microscope was ingenious, but it was difficult to use. Modern microscopes are more sophisticated. The eyepiece lens is set into the top of a tube. At the other end of the tube is a turret of objective lenses. This turret can be revolved to position different objective lenses at the bottom of the tube. Each objec-

162

A young student using a standard optical microscope. With a microscope of this kind you can examine a great variety of small objects, as you can see on the opposite page.

tive lens provides a different degree of magnification. The focus of the microscope is adjusted by two knobs; one is used for large movements, the other is used for fine movements.

The specimen—the object you want to examine—is generally placed on a glass slide, and this is held in place on a frame called the stage by two clips. Light from below is directed on to the object through a hole in the stage. Some microscopes use a mirror to reflect light, which is then focused on the specimen by an arrangement of lenses called a condenser. Other microscopes have their own light source instead of a mirror.

Small specimens, such as hairs and the tiny organisms that live in pond water, can easily be viewed under a microscope. Larger objects need careful preparation. Very thin slices of plant or animal tissue can be made by using an instrument called a microtome. These slices can be mounted on slides and stained with chemicals to show up particular features. Thin slices of rocks can be cut, using a diamond saw, and mounted in transparent plastics material. They can then be ground to the required thickness for viewing under a microscope. Metals are examined under microscopes by shining light on to their surface.

Microscopes that use light can be focused to look at objects as small as 0.0001 mm. Objects smaller than this must be examined using a completely different type of microscope, called an electron microscope. In this a beam of electrons is focused on to the specimen. The beam is altered according to the shape and thickness of the specimen, and the altered beam is then directed on to a television screen, where it forms an image. Using an electron microscope, objects only millionths of a centimetre across can be magnified over 500,000 times.

Specimens for examination in an electron microscope need very careful preparation. For scanning by the electron beam they are placed inside a vacuum chamber because the presence of air would interfere with the beam. Specimens must be exceedingly thin, generally less than one ten-thousandth of a millimetre. They cannot be mounted on glass, but can be laid on a thin film of cellulose nitrate, which is supported on a very fine metal mesh. An electron microscope enables scientists to see substances magnified up to 2,000,000 times, and single atoms can be seen.

Like the ordinary microscope, the electron microscope was the work of many men. Most of the designing was done by a group of German scientists in the 1930s. Its development was very rapid, and within 15 years it became one of science's most useful tools. Improvements are still being made.

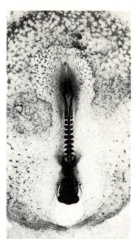

Microscopic pictures:
Left: The mouthparts of a blowfly; Above: The eye of a small mammal; Right: An embryo chick, 36 hours after the egg was laid.

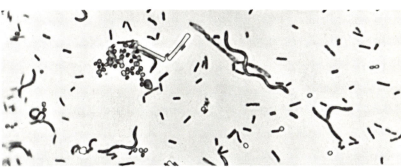

Above: Several kinds of bacteria; Left: A bee's leg, with pollen on it; Below: An adult gnat emerging from the pupa.

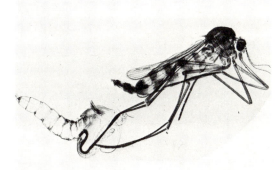

Above: A hydra, a tiny pond animal, with a bud from which a fresh hydra will grow. Right: Cells scraped from the lining of a person's cheek.

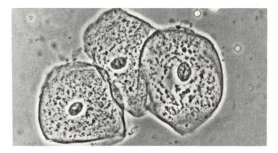

The Mighty Atom

Uncontrolled atomic power: The mushroom cloud produced by an atom bomb, while deadly radiation spreads for a long way around it.

The paper on which this book is printed is made up of millions of tiny particles. Your body, your clothes and in fact all materials are made of such particles, which are called atoms. Each one is about one hundred millionth of a centimetre across, and the thickness of this page is about two million atoms.

An atom consists of a tiny central nucleus, which itself is composed of two kinds of particles—protons and neutrons. Around the nucleus the rest of the atom consists of a cloud of even smaller particles called electrons. These electrons are some distance away from the nucleus, and thus the atom is composed mostly of space!

Each electron has a negative electrical charge and each proton has a positive charge. These charges attract each other and hold the atom together, but a stable atom is not itself electrically charged because the number of electrons is always equal to the number of protons.

There are 106 different kinds of atom known at present, and each kind is the basic building block of an element, a simple chemical substance. The atoms of different elements have different numbers of protons and electrons. For example, a hydrogen atom has one electron and one proton; and an atom of uranium has 92 electrons and 92 protons.

The atoms of most elements can exist by themselves, but some elements, such as hydrogen and oxygen, can only exist as two atoms joined together. A combination of more than one atom is called a molecule. Most of the materials we use are made up of molecules, and they usually contain atoms of more than one element.

For example, a water molecule is composed of one oxygen atom and two hydrogen atoms. The paper on which this book is printed is composed of even larger molecules, which mostly consist of carbon, hydrogen and some oxygen atoms. However, these molecules are still very small. This page is about 100,000 molecules thick.

Why do atoms form molecules? The answer to this lies in the structure of the atom. The electrons that surround the nucleus are arranged into layers, or shells. A hydrogen atom (symbol: H) with its one electron has one shell, but it would be more stable if it had two electrons. Therefore, two hydrogen atoms share their electrons, and this forms the bond of the hydrogen molecule (H_2). An oxygen atom (O) has two shells, and the outer shell contains six electrons, but it would be more stable if it contained eight. Therefore it shares two of its electrons with another oxygen atom, thus forming an oxygen molecule (O_2). In a water molecule (H_2O) the oxygen atom shares one electron with each hydrogen atom, and the hydrogen atoms share their single electrons with the oxygen atom.

Salt, or sodium chloride (NaCl), molecules are formed in a different way. The sodium atom gives one of its electrons to the chlorine atom. The result is that the sodium atom becomes positively charged, and the chlorine atom becomes negatively charged. The two charged atoms are attracted to each other, and the molecule is formed.

An atom of a particular element always contains the same number of protons. For example, all hydrogen atoms have one proton, but sometimes the number of neutrons changes. Atoms of the same element with different numbers of neutrons are called isotopes. For example, hydrogen has three isotopes—hydrogen-1 (with no neutrons), hydrogen-2 (with one neutron), and hydrogen-3 (with two neutrons). These atoms are stable and remain as they are, but the isotopes of some elements are unstable and emit particles and rays. Such isotopes are called radioactive. Examples of elements with radioactive isotopes are radium and uranium.

The discovery of radioactivity led to the production of the first atomic bomb. The bomb contained uranium-235, a rare isotope of uranium. When an atom of uranium-235 absorbs a neutron its nucleus splits up and emits more neutrons and a lot of energy. This process is called nuclear fission (splitting the nucleus of an atom). If the emitted neutrons

are absorbed by other uranium-235 atoms, they too split and a chain reaction begins.

In the ground, where uranium-235 atoms are few and far between, this does not matter, but if enough uranium-235 is collected together, it will exceed the amount called the critical mass. In this case the energy released by the nuclear fission chain reaction causes a violent explosion.

The same process is used, under controlled conditions, in the reactor of a nuclear power station. The nuclear fuel is contained in rods. Neutron-absorbing control rods can be lowered into the reactor to remove neutrons when the reaction has to be slowed down or stopped. Nuclear reactors can use uranium-235 as fuel, but other radioactive isotopes, such as plutonium-239 and uranium-233, are also available. They all produce heat, which can be converted to electricity.

Controlled atomic power: A close-up view of the fuelling machine in a modern nuclear power station, which went into production in the 1970s.

The Story of Plastics

Our ancestors relied entirely on natural materials for their needs. They used wool from animals, cotton and linen from plants and silk from silkworms to make fabrics for clothes and furnishings. Other household items were made from wood, metal, glass or clay. These materials are still very much used today, but we have an additional, man-made group of materials—the plastics. Without these life in our modern society would be very different.

The first plastics material was made by an American printer, John Wesley Hyatt, in the 1800s by treating cellulose from plants with sulphuric and nitric acids. This produced celluloid. In 1908 a Swiss chemist, Jacques Brandenberger, invented cellophane, also made from plant fibres.

Today most plastics are made entirely from man-made chemicals. In 1909 an American chemist, Leo Baekeland, produced Bakelite by mixing together phenol and formaldehyde. This first, truly synthetic plastics led the way for a whole range of new materials.

Plastics get their name from the fact that they *are* plastic—that is, they can be moulded

Above: A range of kitchen food containers in polyethylene. These containers are hard and transparent.

Left: Making Christmas decorations from plastics. The basic polystyrene has been given a shiny metallic coating, ready for the final decoration.

Here are four everyday things made from plastics: A toothbrush, a mug, paint, and a telephone. You can probably think of many more if you take a look round your own home.

or bent into almost any shape. This is usually done by heating them and applying pressure. This very special property of plastics is due to the shape of their molecules. All plastics consist of very large molecules that are made up of long chains of carbon atoms. They are made from substances with short molecules by treating them with chemicals, so that the molecules join up to form long chains.

Once formed, these long molecules can slide over each other, particularly when heated. As a result plastics can be moulded into shapes or drawn out into long threads. However, not all plastics have exactly the same properties, and there are two main groups—thermosetting plastics and thermosoftening plastics.

Thermosetting plastics are hard and difficult to bend. They are also resistant to heat. This group includes urea-formaldehyde resins, melamine-formaldehyde resins, epoxy resins and Bakelite (a phenol-formaldehyde resin). These plastics are used to make such things as light fittings, screw caps for bottles and heat-resistant worktops.

The properties of thermosetting plastics result from the way in which the molecules are arranged. The long-chain molecules are joined together by cross-links, which prevent the molecules from sliding over each other. If a thermosetting plastics is warmed, more cross-links form, making the plastics even harder and heat resistant.

However, thermosetting plastics do melt if they are heated to a high enough temperature. They can therefore be shaped, and the most common method of doing this is by compression moulding. The material, usually in the form of pellets, is placed in the bottom half of a mould. The top of the mould is then clamped down and the mould is heated. The pellets melt and take up the shape of the mould.

Kitchen work surfaces and table mats—sold under trade names such as Formica and Argolite—are made by laminating melamine-formaldehyde resins. Layers of paper are soaked in the resin and laid one on top of another. Then they are clamped in a press and heated.

Thermosoftening plastics, or thermoplastics for short, are softer and melt when warmed. This group contains the best-known plastics. Nylon and polyesters are used to make clothes. Polystyrene, mixed with air bubbles, is used to make insulating material. PVC (polyvinyl chloride) is used to make drain pipes and 'wet look' clothes. Perspex is used as a substitute for glass. Polyurethanes are used in paints and varnishes. PTFE (polytetrafluoroethylene), the most heat-resistant thermoplastic, is used to make non-stick coatings for cooking pans and utensils.

Thermoplastics are much easier to shape than thermosetting plastics. They melt more easily and remain molten for a long time without any chemical change taking place. Polythene is made into buckets and toys by injection moulding. In this process pellets of plastics are melted in a heated chamber and then forced into a cooler mould. The mould can then be opened to remove the moulded object.

Hollow articles, such as polythene bottles, are made by blow moulding. Air is forced into a piece of softened plastics inside a water-cooled mould.

Some thermoplastics are extruded, or forced through a die hole of the required shape. For example, rods are made using a round hole, and a film is made when the die hole is a thin slit. Nylon and polyesters are made into fibres by forcing molten plastics through a spinneret—a special kind of die with many tiny holes.

Right: A cellulose acetate form of plastics takes a hammering: it is very tough. Far right: The non-stick coating on cooking pans is made from PTFE, the most heat-resistant of the thermoplastics.

167

Lasers—Light at Work

Ray guns have been used by science fiction heroes for many years. In the 1964 film *Goldfinger* James Bond came near to being killed with a ray that could cut through gold. At the time experts said that this technique was not yet possible—but four months later the 'death ray' changed from science fiction to science fact.

The first successful laser was built in 1960 by Theodore Maiman, an American scientist. His laser could not cut gold, but it could produce brief pulses of an intense beam of light. The initials of the name 'laser' stand for **L**ight **A**mplification by **S**timulated **E**mission of **R**adiation. When atoms are stimulated in a particular way they can be made to produce light—a form of electromagnetic radiation (see pages 174-175). The name 'laser' thus simply means that this instrument increases the amount of light produced when atoms are stimulated.

Maiman's laser contained a crystal of ruby surrounded by a flash tube. At one end of the crystal there was a silvered mirror, and at the other end there was a half-silvered mirror. In such a laser a flash of light from the flash tube causes some of the ruby atoms to gain energy and become excited. These atoms then lose their energy by emitting light. The light travels down the ruby crystal, but is reflected back by one of the mirrors. It continues to pass up and down the crystal, being reflected off the mirrors. At the same time it excites more ruby atoms, which produce more light. When all the ruby atoms are producing light some of it passes through the half-silvered mirror.

However, the beam of light that emerges is not ordinary light. Your table lamp produces light that is made up of several wavelengths, but when ruby atoms are stimulated they all produce light of the same wavelength. The waves are all in step with each other and the beam of light that emerges from the laser is called coherent light. This beam is narrow and it can travel over very great distances. Even a powerful, concentrated ordinary light, such as a car's headlight, produces a wide beam and carries only a comparatively short distance.

A ruby crystal was the first material to be used in a laser. Many substances are now used, including gases such as argon and neon. Some lasers produce short pulses of very intense light. Others produce continuous beams of light.

Lasers now have many uses because of the very special properties of the light they emit. The straightness of the beam is used in building work for aligning and guiding. The new London Bridge was built using a laser. The

Science fiction becomes science fact: A laser beam is used to cut through a steel plate.

narrowness of the beam has enabled scientists to measure the Earth's distance from the Moon. Laser beams are fired at mirrors left on the Moon by Apollo astronauts. Even at that enormous distance the beam spreads over only three kilometres.

The coherence of laser beams is used to make 3-dimensional pictures called holograms. The purity of laser light is being used to transmit telephone conversations. A beam of laser light travelling down a glass fibre can carry hundreds of telephone conversations at the same time.

Laser beams can also be used to transmit sound in space. Two people at either end of a beam can talk to each other without fear of

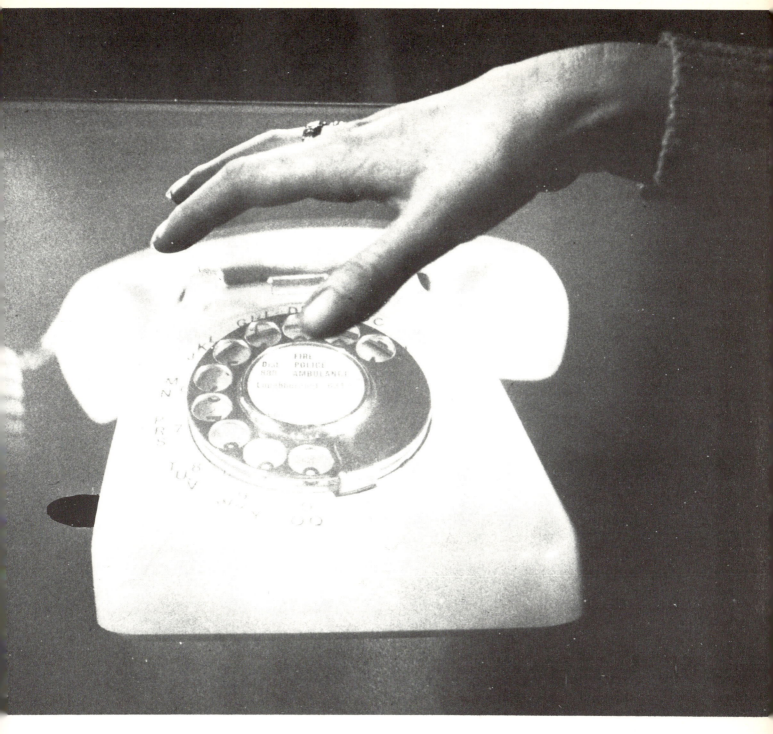

being 'bugged'. Laser beam transmission can be particularly useful in poor weather conditions because laser light cannot be blocked by mist or fog.

The first lasers only produced low-power beams of light, but soon scientists made a carbon dioxide laser which produced a beam so intense that it created heat and could cut through steel. Today such lasers are used to cut all kinds of metals, jewels and even piles of cloth for making mass-produced suits. Cutting with lasers can be done very accurately. They are used to make tiny holes in the teats of babies' bottles and also to make incisions in surgical operations. A laser can also be focused on the back of the eye to 'pin

back' pieces of detached retina—the light-receiving layer that lines the inside of the eye.

There are also military applications for lasers. Portable 'death ray' guns are not yet possible because of the bulky equipment needed to produce such power, but there are laser rifle sights and laser-guided missiles. It is also claimed that lasers have been used to 'knock out' spying space satellites.

The possible uses for lasers have not yet come to an end. One of the most exciting developments is the invention of holograms, laser photographs. An image with laser light is recorded on a plate. When the plate is viewed with laser light, a three dimensional image is formed.

The hand is real but the telephone is not: it is a hologram which is produced by a number of laser beams. Also, in the future, telephone conversations may be transmitted by beams of laser light.

Using Metals

There are 92 elements that occur naturally in the Earth's crust, and most of these are metals. Metals are substances that are good conductors of electricity and heat. Some metals, such as aluminium, iron and copper, are very common and occur in large amounts. Others are more difficult to find and extract. Gold and platinum occur in the ground as free metals, but most metals occur as ores. These are chemical compounds of metals with other elements, such as oxygen and sulphur.

The earliest prospectors used picks and shovels to search for ores, but today there are more advanced methods of finding metals. Geologists use instruments called gravimeters to detect different densities in the rocks below. Magnetometers are used to detect

magnetic minerals, such as the ores of iron and cobalt. Radioactive metals, such as uranium, are found by using geiger counters.

After a metal ore has been found, it is extracted from the ground by mining. Then several stages may be needed to obtain the pure metal. The ore is usually crushed first, and if possible the metal-containing particles are separated from unwanted rock.

Copper and nickel ores are mixed with water and the metal sulphides are separated from the earthy material, or gangue, by flotation. The sulphides float to the surface as a froth. The concentrated ore is then roasted in air. This removes other unwanted chemicals, such as sulphur, and fuses the particles into more manageable lumps.

Iron ores occur mainly as iron oxide (iron combined with oxygen). This is washed to

The sparks fly as molten metal is poured from a giant ladle at a modern steelworks. Steel is made at temperatures of more than 1,600°C.

remove the gangue and then roasted. The concentrated ore is then heated to a high temperature in a blast furnace by burning coke. The molten iron sinks to the bottom of the furnace, from where it is periodically removed. Limestone is also added to the furnace. This helps to remove the impurities, forming slag which floats on top of the iron. The ores of tin and lead are also heated in furnaces to extract the metals.

Some ores are treated with acids or other substances to remove the metal as a dissolved compound. Copper oxide ores are treated with sulphuric acid to produce copper sulphate. Gold is sometimes removed from its ore by the use of cyanide.

A commonly-used method in extracting and refining metals is electrolysis. An electric current is passed between two electrodes through a solution of the metal ore or through the molten ore. The metal is deposited on the negative electrode. Sodium metal is extracted from salt (sodium chloride) in this way.

Metals can be cast in moulds, hammered into shapes, extruded through holes and electroplated on to the surface of other metals. They have a tremendous variety of uses. Frequently one metal by itself does not have all the required properties and so a number of metals are combined to form alloys.

Brass is an alloy of copper and zinc, and sometimes a little tin. Bronze consists mostly of copper, alloyed with tin, aluminium or manganese. Pewter is an alloy of lead, antimony and copper. Steel is an alloy of iron and carbon, together with small amounts of other metals, such as manganese, vanadium, molybdenum and tungsten. Stainless steel contains nickel and chromium as well.

Steel is a very important alloy because of its hardness. It is made from iron by removing most of the carbon and other impurities. This is done by heating the iron in a special type of furnace. Carbon steels contain about 0.08 per cent to 1.5 per cent carbon in the form of iron carbide, or cementite, and their properties depend on the amount of carbon.

Other metals have a wide range of uses. Aluminium is used in cooking pans, aircraft and engines. Copper is used in electricity cables. Gold, silver and platinum are all used in jewellery. Silver is also used in photography, and gold is used in dentistry. Lead is used in plumbing and car batteries.

Mercury, the only metal liquid at normal temperatures, is used in thermometers. Sodium and potassium are used as compounds in fertilisers. Zinc is used as a coating on iron and steel to protect them from corrosion—a process known as galvanising. Tungsten, the metal with the highest melting point, is used for the elements of electric-light bulbs.

Extracting iron ore in Mauritania. In the upper inset picture you can see the terraces, like giant steps, cut in the mining, which is by the opencast system. The lower inset shows an aerial conveyor belt carrying the ore.

Making Textiles

Until the end of the 1800s the only textiles available were cotton, linen, wool and silk. All of these are natural fibres. As early as 1664 the English scientist Robert Hooke watched silkworms at work and predicted that one day man would be able to imitate them. Over 250 years later he was proved right, and today we have a wide range of man-made textiles.

Cotton fibres are found in the bolls (seed pods) of cotton plants, which mostly grow in the tropical areas of the world. The cotton bolls are harvested and they then undergo a process called ginning, which removes seeds, pieces of leaf and other impurities. Then the fibres are spun into yarn.

Linen is made from the fibres of flax plants. The fibres grow in the stems of the plants. The harvested plants are subjected to a series of treatments. During retting they are left in a damp place to allow bacteria to destroy all the soft parts, leaving the fibres and the wood. The plants are then scutched to remove the wood, which is first broken up by rollers and then scraped away. Finally the fibres are combed in a process called hackling. This removes short fibres and arranges the long fibres ready for spinning into yarn.

Wool is animal hair. Camel's, goat's and rabbit's wool are all used, but the most common is sheep's wool. After wool has been sheared (cut off the animal) it goes through several processes. It is first sorted and graded, and the best quality hair is kept for the finest fabrics. Then it is thoroughly washed and scoured to remove grease and dirt. Pieces of plant material are removed by a mild acid.

Next the wool is carded, drawn between spiked rollers to straighten out the fibres. Before being spun into yarn the wool is

The spinning wheel, which was invented in the Middle Ages, was used for hundreds of years for making thread.

A modern loom weaving towels for use in a hospital. The hospital's initials are incorporated in the design.

combed to remove the shorter fibres. Worsted yarn is made from fibres 6-13 cm or more in length. Woollen yarn is made from fibres 3-8 cm in length.

Silk is the finest natural fibre of all, and historians believe the Chinese used it as long ago as 2500 BC. It is made by the silkworm, which is the larva of the silk moth. Before changing into moths silkworms build themselves cocoons made entirely of silk filaments. The cocoons are collected, and the filaments are carefully unwound by skilled workers. Several filaments are twisted together to make the silk threads used in weaving.

The first man-made fibre produced commercially was called 'artificial silk'. It was literally an attempt to copy the silkworms. In 1889 Count Hilaire de Chardonnet, a French chemist, used plant material, which is composed of cellulose, to make cellulose nitrate. He dissolved this substance in a mixture of alcohol and ether, and it formed a sticky liquid which could be drawn out into threads. To mass-produce the threads, Chardonnet used a metal plate pierced with holes—called a spinneret—through which the threads were drawn.

Today many artificial fibres are made of cellulose. During the manufacturing process the cellulose may be changed into another chemical and then back into cellulose. Exam-

American Indians on a farm near Cusco in Peru weaving cloth on a primitive loom.

Below: Cotton bolls (seed pods) ready for picking.
Bottom: An early Jacquard loom. This loom, invented in 1801 by a French weaver, Joseph Jacquard, uses punched cards to control the loom mechanism and so weave patterns automatically.

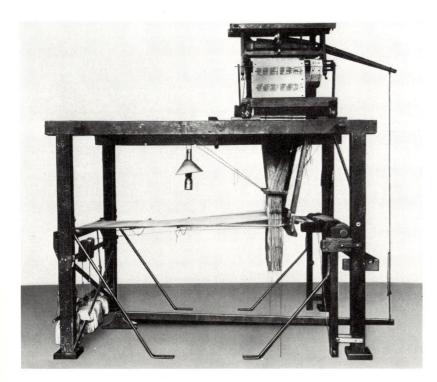

ples of such regenerated cellulose fibres are cuprammonium rayon and viscose rayon. Acetate rayon is made from cellulose acetate.

Completely synthetic man-made fibres are special kinds of thermoplastics (see pages 166–167). The long-chain molecules of which they are made enable these plastics to be drawn out into long threads. Synthetic fibres include polyamides (such as nylon), polyesters (such as Terylene, Dacron and Crimplene) and acrylic fibres (such as Orlon, Acrilan and Courtelle).

The details of the manufacture of synthetic fibres vary according to the chemicals involved, but the basic methods are the same. In the manufacture of nylon, adipic acid and hexamethylene diamine are reacted together to form nylon salt. Crystals of nylon salt are washed and dissolved in acetic acid. This mixture is heated for two hours at 250°C to concentrate the solution. Then the temperature is raised to allow the long-chain molecules of nylon to form. The nylon is then cut into chips and melted in the top of a spinneret. It is forced through the holes, and long nylon threads are formed.

Nylon fabric is often used instead of cotton or linen. It is a stretchy fabric and hence is used to make tights and stockings. In sheets and shirts it has the advantage that it will not crease. However, a mixture of Terylene and wool gives a less glassy appearance, and Terylene holds heat-set creases more or less permanently. Acrylic fibres are often used instead of wool, with the advantage that they do not shrink.

Energy and Waves

The word 'energy' is used very often today. Foods are described as being 'full of energy', and people talk of an 'energy crisis', but the word has only one meaning; it is simply the capacity to do work.

A raised hammer has potential energy because of its position. As the hammer falls this potential energy is converted into kinetic (moving) energy, which is converted into heat energy as the hammer strikes an object. Electrical energy generated at a power station does work when it is converted into heat and light in the home.

Energy can also be transmitted in the form of waves. For example, if you drop a stone into a calm pond, the kinetic energy of the falling stone is transmitted across the pond as waves. The waves form a series of regular crests and troughs, and the distance of one wavecrest to the next is called the wavelength.

Sound is transmitted through the air as waves, but the waves are formed in a slightly different way. The 'crests' are areas where the molecules of air are pressed together (compressions), and the 'troughs' are areas where the molecules are farther apart (rarefactions). Thus, sound waves from a loudspeaker reach the ear as a succession of compressions and rarefactions. The pitch of the note depends on the frequency (the number of compressions that reach the ear in a particular time). Sound always travels through the air at about the same speed (340 metres a second). Hence, if the wavelength is short the frequency is high.

These principles also apply to another form of travelling energy, called radiant energy. This can travel through space in the form of electromagnetic waves. These all travel at the same speed (299,792,500 metres a second). The range of electromagnetic waves is called the electromagnetic spectrum. Starting with the longest wavelengths, this spectrum consists of radio waves, infra-red rays, visible light, ultra-violet rays, X-rays, gamma rays and cosmic rays.

We use the longer radio waves to transmit messages round the world and into space. Television signals are carried by shorter radio waves. The shortest radio waves are microwaves, which include the radar waves used to detect ships and aeroplanes.

We feel infra-red rays as heat, and the Sun's heat reaches us in this form. The heat felt in front of an electric radiator is also the result of infra-red rays. In fact most objects give off infra-red rays, and photographs taken with film sensitive to these wavelengths can reveal things that are otherwise invisible.

Between the wavelengths of 7.6×10^{-7} metres and 4×10^{-7} metres there is a band of electromagnetic waves that you can see. This visible light is divided into colours. Red is at the lower end of the visible spectrum and violet is at the upper end. You see the combination of all these colours as white light, but they can be split up by using a prism, a wedge of glass.

Light has very useful properties. It is reflected by all objects, but very flat surfaces do not distort it. Mirrors have extremely flat reflecting surfaces that allow you to see images. Light can also be bent by transparent objects, such as glass. This property is used to create magnified images in magnifying glasses and telescopes.

Above the visible spectrum are ultra-violet rays. Although they are invisible to human beings, some insects such as bees can detect

Below: A stone thrown into a pond creates waves as its energy spreads across the surface of the water.

them. Flowers that are white to us do not appear white to them. Ultra-violet rays from the Sun can cause sunburn.

Waves with the shortest wavelength have the highest energy. X-rays pass through some parts of the body but are absorbed by others, such as bones. Thus they can be used to make X-ray photographs. Gamma rays have the same properties but they are far more penetrating and are too dangerous to use. Some radioactive materials give off gamma rays.

Cosmic rays are the most energetic of all electromagnetic waves. Cosmic rays from the Sun cause problems for astronauts in space, but they have been used to make 'cosmic ray pictures' of such structures as the Great Pyramid of Pharaoh Khufu in Egypt.

Opposite: X-rays are a form of waves that have great value in medicine. This small child is being examined for possible tuberculosis at a mobile X-ray unit in Lesotho, run by the World Health Organisation.

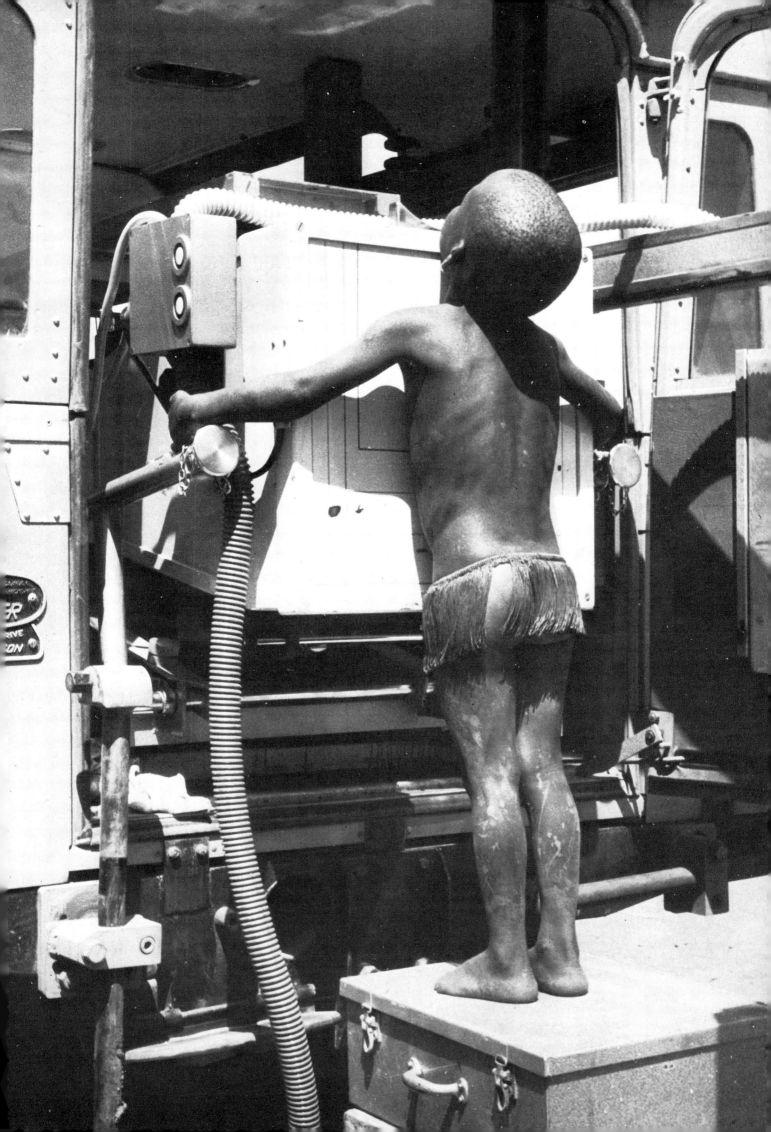

Science Inquiry Desk —2

What makes a rainbow?

You see a rainbow when the Sun is shining from behind you, and rain is falling in front of you. The white light that comes from the Sun is a mixture of all the colours, but if it is passed through a prism, a wedge-shaped piece of glass, the rays of light are bent and it splits into all the colours that form what is called the spectrum.

A rainbow is formed because each drop of rain acts as a miniature prism. It has two bands of colours. The brighter, inner arch has red on the outside and violet on the inside. The fainter, outer arch has the order of colours reversed.

? ? ?

Do trains have to run on rails?

Trains need rails to guide them—but they do not necessarily have to run along them on wheels. Several new kinds of trains are being developed that have no wheels.

French engineers have built the 'Aerotrain'. It runs along a raised track made of concrete and shaped like the letter 'T' upside down. When it is still, the train sits on the cross-piece of the 'T' and the upright fits into a deep slot in the train. In motion, the train rides clear of the track on a cushion of air, with the upright guiding it. It is driven along by a propeller at the rear.

The magnetic levitation train, called 'maglev' for short, is being developed by the Japanese. A motor inside the train sets up a magnetic field, just like the magnetic field that makes an electric motor spin round. The magnetic field acts against the track, lifting the train clear of the rails and pulling it along.

? ? ?

Why does a spoon in a glass of water look as if it is bent? (See left).

Light travels at different speeds in different substances, even though they all seem to be transparent. So when it passes at an angle from one substance to another it is refracted (bent). Because it enters the glass from the air at an angle part of the light-wave enters the water before the rest and is slowed down. The other part of the light-wave continues at the same speed until it too strikes the water. As a result the light-ray bends.

? ? ?

Is rubber one of the plastics?

Not really—though it is similar to plastics because it is made up of long chains of molecules, just as plastics are. Unlike most plastics, rubber is elastic. It can be bent or stretched, but then springs back to its original shape when released. Also, rubber is a natural product. It is made from the milky-looking sap, called latex, which comes from rubber trees.

Today chemists make synthetic (artificial) forms of rubber, mostly from petroleum. Most of them are

very similar to plastics in their chemical make-up.

? ? ?

Who invented the telescope?

Nobody knows for certain who invented the telescope, but we do know it happened in the Netherlands in 1608. An old story says the inventor was a spectacle maker named Hans Lippershey. One day he was holding a spectacle lens in each hand, and happened to look through both of them at once at a nearby church tower—and found that the tower seemed to be nearer. So he mounted the lenses in a tube, and the telescope came into being.

The great astronomer Galileo Galilei heard of Lippershey's discovery and promptly made his own telescope. With it he discovered the mountains on the Moon.

? ? ?

Is quicksilver really silver?

No, quicksilver is an old name for mercury, the only metal which is liquid at normal temperatures. An old meaning of 'quick' is 'living', and mercury seems to be almost alive. If some is spilled it forms into little balls ballast tanks. To come to the surface again, the submariners blow the water out of the ballast tanks with compressed air, and the boat rises. Fins like those of a fish help control a diving boat.

? ? ?

What makes heat travel along an iron bar?

As you can see on pages 174–175, heat travels through space in the form of electromagnetic waves. Heat can travel in two other ways, called conduction and convection. The heat in an iron bar travels by conduction.

When you heat one end of the bar, as when you put a poker in the fire, the molecules at that end of the bar start to move about rapidly. They strike the molecules next to them, so that they too start to move more quickly and grow hot. In time, the movement and the heat travel down the whole bar. Iron and most other metals are good conductors of heat, but many substances such as wood and bone are bad conductors.

In convection, heated matter actually moves from one place to another. Convection occurs in liquids and gases. Hot air rises, which is why gliders are able to soar high in the sky, riding on a column of warm air called a thermal. You can see convection clearly if you look at the bubbles rising in a pan of boiling water.

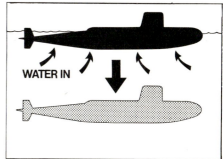

How a submarine goes up and down.

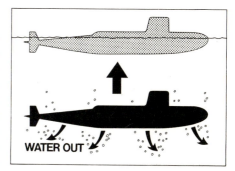

of liquid which run about. The reason why it is also called mercury is that Mercury, in old Greek legend, was the fleet-footed messenger who ran errands for the gods.

? ? ?

Why does metal go rusty?

Rust affects only iron and steel (which is made from iron). It is caused when oxygen from the air combines with some of the iron in a process called oxidation. The iron and the oxygen form a compound known as iron oxide. If the metal is damp, the iron oxide combines with water to form rust, a reddish-brown substance. Rust destroys the surface of the iron, and in time will eat its way through it.

Stainless steel does not rust because it contains the metal chromium which resists oxidation.

? ? ?

How does a submarine stay under the water and then come up again?

Things float in water because they are lighter than water. Although the hull of a submarine is made of steel, which is heavier than water, the whole boat,

Above: Rust has eaten away at these scissors.

like all steel boats, floats because it is full of air, which is much lighter than water. To sink, a submarine lets seawater into tanks called ballast tanks inside the hull, until the boat weighs more than water. The heavier the boat the deeper it sinks, so submariners control the depth of a dive by the amount of water they let into the

177

Rolling Along the Road

Until the 1800s travel by road was very slow. A journey of 100 kilometres could take two days by horse-drawn carriage. Today such a journey may take less than two hours, and this is due almost entirely to the internal combustion engine.

The name 'internal combustion engine' describes exactly what it does. Combustion means burning, and fuel is burned inside the engine in chambers called cylinders. The burning of the fuel produces gases that expand

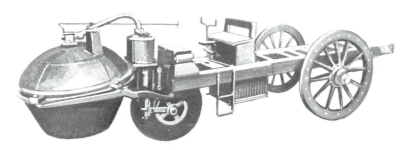

The world's first engine-driven road vehicle: a steam gun carriage built by Nicolas Cugnot, a French army captain, in 1770. It travelled at 4 kph, and had to stop every 15 minutes for more water.

violently. The energy of this expansion is used to drive the vehicle.

The petrol engine of a modern car has four or more cylinders and each one works on what is called the four-stroke cycle. On the first stroke, or induction stroke, the piston inside the cylinder moves downwards and a mixture of air and vaporised fuel is drawn in through the inlet valve. The second stroke is called the compression stroke. The piston moves upwards, compressing the gases into a small space inside the cylinder. At the beginning of the third stroke, or power stroke, the sparking plug ignites the compressed gases. The burning gases expand, forcing the piston down and providing the driving force which the engine produces. On the fourth stroke, or exhaust stroke, the piston rises again, the exhaust valve opens and the waste gases are pushed out of the cylinder.

A diesel engine is another kind of internal combustion engine. Like the petrol engine it

An Italian F.I.A.T. car built in 1899. It had two cylinders mounted horizontally (most modern cars have their cylinders mounted vertically).

works on a four-stroke cycle. However, the ignition of the fuel is achieved in a different way. The induction stroke draws only air in through the inlet valve. During the compression stroke the air heats up as it is compressed. (You have probably noticed how a bicycle pump gets hot after the air has been repeatedly compressed in it). At the end of the compression stroke fuel is injected into the hot air. It ignites immediately, resulting in the power stroke of the piston.

A car is started by using electric current from the battery. When the ignition switch is turned the battery operates the starter motor. This engages the flywheel and turns the engine, resulting in fuel being drawn into the cylinders. With the help of the coil—a device which increases the voltage—and contact-breakers the battery also produces the necessary sparks across the sparking plugs. After

The famous 'Model T' car which Henry Ford produced for 19 years, from 1908 to 1927. The car was designed for the ordinary motorist, and 15,007,033 Model T's were built. The car was popularly known as the 'Tin Lizzie'.

the engine has started, the generator, or dynamo, takes on the rôle of the battery in producing sparks, and also keeps the battery fully charged.

The engine has many moving parts and these have to be kept lubricated with oil. This is constantly circulated round the engine from the sump—the reservoir of oil underneath the engine. At the same time the burning fuel causes the engine to heat up. If the temperature becomes too high the engine seizes up. Therefore there is a cooling system. In many cars this consists of a jacket around the engine connected by hoses to a radiator. The system is filled with water, which is constantly pumped round. Air passing through the radiator cools the water down before it returns to the jacket. Some cars, however, are cooled by air alone.

Inside the engine the up-and-down movement of the pistons is converted into rotating movement in the crankshaft. This movement

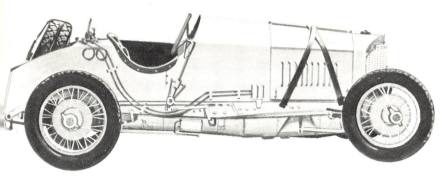

is transmitted through the clutch to the gear-box. The clutch is used to disconnect the gears from the engine so that the required gear can be selected. When the clutch is released the crankshaft turns the gears, which turn the main transmission. This drives the wheels of the car. Some cars have rear-wheel drive and others have front-wheel drive. Cars used in particularly rough conditions have transmissions that can be made to drive all four wheels at the same time.

The other main systems of the car are the fuel system, exhaust system, electrical system, braking system and steering. Petrol reaches the engine from the fuel tank through the carburettor. This device vaporises the petrol and mixes it with air. The accelerator pedal operates a flap called the throttle in the carburettor and this controls the amount of fuel entering the cylinders. The waste gases pass down the exhaust pipe, which is fitted with a silencer to reduce noise.

A 4½ litre Mercedes built in 1914. In that year the cars took first, second and third places in the French Grand Prix race.

Right: The 'Mini' was first made in Britain in 1959, and has been in production ever since. It is light and very cheap to run, using little fuel.

Below: A modern Formula I racing car. Note the super streamlined body and the wing, or airfoil, at the rear which helps to hold the car on the road when it is travelling at high speed.

The electrical system operates the lights, indicators, horn, windscreen wipers and any accessories, such as a demister or a radio. The braking system consists of a number of fluid-filled pipes connected to the brakes. Foot-pressure on the brake pedal is transmitted to the brakes by the fluid. The steering system consists of several rods and linkages that connect the front wheels to the steering wheel.

Petrol-driven cars are noisy, wasteful of energy and create pollution. Many people have studied electric cars as an alternative. However, recent research has shown that electric cars are likely to waste more energy than petrol-driven cars. In addition, electric cars have a maximum speed of about 65 kilometres an hour at present, and they have a limited range. Once the batteries have been used up they must either be exchanged or recharged. Recharging takes too long and battery-exchange stations are too expensive. Thus it will be a long time before the petrol-driven car is replaced by the electric car.

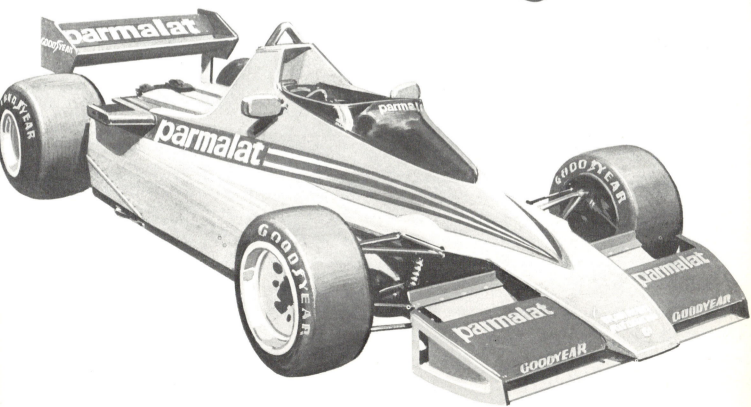

Travelling by Train

In the 1700s, during the Industrial Revolution, people found it necessary to transport large quantities of goods. Thus the railways were born. Railways were not a sudden invention. They evolved gradually over three centuries.

In the 1600s wooden rails were being used in mines to guide trucks. Iron wheels were introduced in the 1700s. These quickly wore out wooden rails, and so iron rails were made. The early trains were pulled by horses, but the invention of the steam engine in England revolutionised the railways. Gradually trains became faster and faster. As a result more efficient methods of braking and signalling had to be developed. Thus the modern railways are the result of several hundred years of experience.

Today the railways of the world cover 1,207,000 kilometres. Some trains are still pulled by steam locomotives, which get their power by burning coal, but most modern trains are pulled by electric or diesel-electric locomotives. Electric trains get their power from electrified overhead cables or rails. In a diesel-electric locomotive the diesel engine powers generators. These provide electrical power for traction motors, which turn the wheels. This system is used, rather than direct drive from the diesel motors, because the electric motors provide better and smoother acceleration.

The main difference between trains and other forms of land transport (except for trams) is that they run on two-rail tracks and do not have to be steered. The tracks consist of two steel rails fastened to wood, metal or concrete sleepers (cross-ties). Railways today use basically the same kind of rail that was used in the 1830s. Since then much research has been carried out to make the metal of which rails are made stand up to the pounding of heavy locomotives and carriages.

The steel wheels of locomotives, carriages and goods wagons have flanges on the inside. The two flanges of a pair of wheels hold them firmly on the track. Points (sections of rail which move) are used to enable trains to switch from one track to another.

The distance between the wheels, called the gauge, varies a lot. More than half the world's tracks are laid in what is known as standard gauge, which is 4 feet 8½ inches (143.51 centimetres). This was the width of tracks at some collieries on the first railway line ever built, the Stockton and Darlington line in northern England, and was adopted so that existing wagons could run on the new lines. Standard gauge is used in North America and all of Europe except the Soviet Union, Fin-

land, Spain, Portugal and Ireland. It is also used in China. Central America, South America, Africa, Australia and other parts of Asia use a variety of gauges, from narrow to broad. The number of different gauges means that trains in these regions cannot run from one railway line on to another, so through journeys are impossible.

The length of rails varies also. In the United States a standard rail is 39 metres long, while in Europe it is 30 metres. On some tracks the rails are welded together as they are laid for smoother travel. Welded rails may be up to 0.4 kilometres long.

The ground beneath the track is called the roadbed. Before the track is laid this roadbed must be carefully levelled. The grade, or gradient, of the roadbed must not be too

Above: George Stephenson's locomotive *Rocket*, built in 1829, won a competition sponsored by the Liverpool and Manchester Railway.

Below: The high-speed Turbo-Train of French Railways, developed in the 1970s. It can reach speeds of up to 300 kph.

steep, otherwise the wheels of the locomotive will slip. Where necessary, therefore, cuttings and tunnels are made through hillsides, and embankments and bridges are built over low ground.

When the roadbed has been prepared the sleepers are laid and ballast—crushed rock, slag or volcanic ash—is placed around the sleepers. This allows surface water to drain away from the roadbed. The track is then aligned and the ballast is compacted around and underneath the sleepers. Finally the rails are laid on the sleepers and fixed down with tie-plates. The rails are joined together with joint bars or fish-plates.

Tracks have to be constantly maintained. This involves removing and inserting ballast, fixing rails, tightening the bolts on fish-plates, and making sure the track is aligned. All these things can now be done with machines operated by a small number of men. Electronic inspection equipment is used to detect faults in the track.

The control of railways has become very complicated, due to the large numbers of trains in operation. In particular, an efficient signalling system is necessary to prevent trains from colliding with each other.

Modern signalling equipment uses both semaphore and light signals in an automatic block signal system. The track is divided into a series of blocks. A train entering a block

The Union Pacific engine 'Big Boy' was built in 1941 and was the largest locomotive of its day. It had 24 wheels and burned 22 tonnes of coal an hour at full speed.

causes the signal at the beginning of that block to turn to red. The block immediately behind has an amber signal, and the block behind that has a green signal. The signals are operated by track circuits—electric currents flowing through the rails. A train entering the block short circuits the track circuit and the signal behind turns to red.

The remaining signals and point switches are operated by men in a central traffic control. They have an electrical diagram, also operated by track circuits, that tells them where all the trains are.

Flying Through the Air

Ever since civilisation began people have dreamed of being able to imitate the birds by flying. Many attempts were made, often using weird-looking contraptions. Some of the results were hilarious, others were tragic; but these pioneers led the way for the first aeroplane, built by two American engineers, brothers Wilbur and Orville Wright. Its first flights, which took place in 1903, covered distances of up to 230 metres.

However, this was not the first time that men had taken to the air. In 1783 the brothers Joseph and Jacques Montgolfier, two French papermakers, took off in their hot-air balloon.

Right: Helicopters are used for all kinds of short journeys. Because they can hover over one spot they are useful for aerial patrol work.

They had discovered that if they lit a fire under the open end of the balloon it would rise into the air. This was because hot air is lighter than cool air. Today hot-air ballooning is a popular sport. However, a modern balloon takes its fire with it. A large flame, fuelled by bottled gas, heats up the air at the opening.

Such balloons can fly only where the wind takes them. In 1852 the French engineer Henri Giffard added a propeller to an enclosed balloon filled with hydrogen—a gas that is lighter than air. The propeller was powered by a petrol engine. Giffard's machine was the first airship. Many airships were built following this example. If the wind was not too strong they could go where they liked. The German Zeppelins were enormous airships with the gas contained in rigid aluminium shells. Air-

A Royal Australian Air Force F111 fighter. This front-line defence weapon can fly at up to 2½ times the speed of sound, and has a very long range without refuelling.

ships had two main disadvantages. First, they were very slow. Second, and more important, the hydrogen or coal gas with which they were filled was highly inflammable. After several disasters, including the wrecking of the British *R101* in 1930 (48 killed) and the German *Hindenburg* in 1937 (36 killed), airships were virtually abandoned. Since then the non-inflammable gas helium has been used in airships, but they are now very expensive to build and operate.

By this time, however, aeroplanes had become faster and more efficient than airships. The principles of flight were becoming well known and designers were concentrating on how to improve the speed and performance of aeroplanes.

The principles by which all aeroplanes fly are the same. The cross-section of a wing is shaped in a special way. The distance from the leading edge to the trailing edge is greater over the top of the wing than along the underside of the wing. As a result the air flows faster over the top of the wing. This lowers the pressure above the wing. The higher pressure below the wing gives it lift, and it therefore rises. The amount of lift created depends on the speed of the aeroplane and the angle at which the wing meets the air—known as the angle of attack. If the angle of attack is too great and the speed is too low, the aeroplane stalls (falls, out of control). This is because there is insufficient lift to overcome the weight of the aeroplane.

The aeroplane is controlled by the ailerons on the wings, and the rudder and elevators on the tailplane. The ailerons work by increasing or decreasing the lift on the wings, causing them to rise or fall. If the pilot wishes to turn to starboard (right), he raises the starboard aileron and lowers the port (left) aileron. This causes the starboard wing to drop and the port wing to rise. At the same time he moves the rudder to starboard and the tail swings to port. The elevators are used to control vertical

direction—enabling the aircraft to fly level, dive or climb.

Forward propulsion is provided by the engine-driven propellers or by jets. These propulsion units have to create sufficient thrust to overcome the slowing-down effect—called drag—as the aeroplane moves through the air. The Wright brothers' aeroplane was driven by petrol engines and it had pusher propellers behind the wings. Later aeroplane designers discovered that puller propellers at the front of an aeroplane were more efficient. Gradually, faster and faster speeds were achieved. The jet engine, invented in 1937, has enabled some aeroplanes to travel faster than the speed of sound, a speed known as *Mach 1*. This is about 1,230 kph.

Conventional aircraft need a long runway for take-off and landing. Helicopters and VTOL (Vertical Take-Off and Landing) aero-

One of the world's most modern airliners, the Anglo-French Concorde. This aircraft flies across the Atlantic Ocean every day in 3½ hours, well above the speed of sound. Inset: prelude to the first aeroplane flight—the Wright Brothers with their glider, on which they based their first powered aeroplane.

planes do not need a runway. A helicopter has a large rotor above the fuselage. The blades of the rotor are shaped like wings, and they create lift in the same way as wings when the rotor is revolving at high speed. A VTOL aircraft, such as the Hawker Siddeley Harrier, has jets that can be pointed downwards to enable it to take off and hover. When the jets face back the aircraft moves forward.

Using engines, man achieved his ambition to fly in 1903, but man-powered flight was more difficult. Since 1935 only 20 short man-powered flights have been made. The greatest achievement was that of *Gossamer Condor*, designed and built by the American Paul MacCready. In 1977 it flew for over a mile (1.6 km). Its wingspan was 29 metres, but it weighed only 31 kilogrammes—light enough to take off and fly with the pilot pedalling furiously inside.

Hydrofoils and Hovercraft

Ever since the invention of the internal combustion engine (see pages 178-179), engineers have been studying ways of creating faster and better means of transport. The invention of the hydrofoil achieved the ambition of high speed travel over water, while the hovercraft is the most versatile form of transport that we have because it can go almost anywhere.

A stationary hydrofoil looks like a conventional power boat, with a hull that floats in the water. However, below the water line there are foils. These foils have the same cross-sectional shape as the wing of an aeroplane, and they work in exactly the same way (see pages 182-183). As the hydrofoil gathers speed the underwater foils begin to generate lift. The hydrofoil rises up, and soon the hull is completely clear of the water.

A conventional power boat is slowed down by the water as it moves along, but the design

Below left: A hydrofoil boat with surface-piercing foils. The two foils, fore and aft, are V-shaped, and only part is submerged.

Below right: A hydrofoil boat propelled by water-jets. It rides on three submerged foils, attached to the underside of the hull by vertical struts.

conventional propellers, often mounted on the foils, or by water jets.

The hovercraft is an even more remarkable invention, as it works on a completely new principle. A hovercraft rides on a cushion of air. As a result it can travel over both water and land, and can move easily from one to the other. The idea of the hovercraft was first thought of by a British engineer, Sir Christopher Cockerell, in 1954. In an experiment using two coffee tins, one inside the other, and a hairdryer he showed that air pressure could be used to lift weights. By 1955 he had designed the first prototype; and the first full-scale hovercraft, the SRN 1, was built in 1959.

Today there is a variety of hovercraft designs, but they all work on the same principle. Air is sucked in through large fans at the top of the craft. Around the hull there is a flexible skirt, and the air from the fans is

of a hydrofoil overcomes this problem. Only the foils are moving through the water, and thus the slowing down effect, or drag, is considerably reduced. As a result the hydrofoil can move very swiftly. Some can reach speeds of 50 knots (93 kph).

The first successful hydrofoil was built in 1906, when an Italian engineer, Enrico Forlanini, built what he called a 'hydro-aeroplane'. It had a canoe-like hull, underneath which were several groups of foils arranged like ladders. The craft was propelled by two aeroplane propellers, one at the front and one at the back. It was not until the 1920s that hydrofoils were taken seriously. Since then, much research has been done on the design and positioning of the foils and on methods of propulsion.

At the present time there are hydrofoil speedboats, gunboats, passenger carriers and even catamarans. Propulsion is provided by

pushed down between the skirt and the hull. The pressure of the air as it emerges from the bottom both creates and maintains a cushion of air underneath the hull.

There are variations of this system. Very light hovercraft do not need a double skin. The air is pushed down directly into the hull, which is a simple open box. The sides of the box may be solid or flexible. Another variation is one in which both the skirt and the sides of the hull are flexible. This arrangement automatically adjusts itself to the size of the waves, thus giving the hovercraft a smoother ride over the sea.

The forward propulsion of the hovercraft is provided by pusher propellers, which push air

The British hovercraft *The Princess Anne* roars into harbour after a trip across the English Channel, carrying both passengers and cars. In North America hovercraft are known as ACVs (Air Cushion Vehicles).

backwards. Rudders steer the craft by deflecting the airstream from the propellers. Using this system hovercraft can travel very fast; some can reach speeds of over 150kph.

Hovercraft are so versatile that they now have a wide range of uses. Hovercraft sea ferries are a familiar sight in many places. Hovercraft can also travel over almost any kind of land surface. They have been used for transporting heavy loads across desert areas and even as oil drilling platforms in swampy or icebound places.

Hovertrains are also being designed. A hovertrain runs along a track, but is separated from it by an air cushion. So far only experimental models have been built.

185

Travelling Through Space

Man's journeys in space have all taken place in very recent times. The first manned space flight was in 1961 (see pages 24-25). Space travel, for years only a science fiction writer's dream, has been made possible by the exciting developments in science and technology that have come about since the end of World War II in 1945. One of the most important of these developments has been the growth of electronics, and in particular the invention of the transistor in 1948. By using miniaturised electronic equipment technicians have been able to make spacecraft light enough to send into space.

The main aid to exploring space is the rocket. American scientist Robert Goddard built the first rocket in 1926. It travelled a distance of 56 metres at 104 kph. By 1935 Goddard had built a more powerful rocket that could climb nearly 2·5 kilometres into the air at over 1,120 kph. Since then rockets have become more and more powerful.

Rockets work by generating thrust. This, like weight, can be measured in kilogrammes. A space rocket creates a thrust greater than the total weight of the spacecraft. The extra thrust accelerates, or speeds up, the spacecraft to the velocity it needs for its journey.

A single rocket cannot produce enough thrust to push a spacecraft into orbit around the Earth or send it clear of the Earth on a journey to the Moon or the planets. Therefore three-stage rockets are used. As each stage runs out of fuel it is released. The next stage ignites and continues to accelerate the rocket away from the ground. After the last stage is released only the spacecraft itself remains in orbit above the Earth. It has small rockets which are used for changing direction in space.

When the spacecraft returns to Earth it faces different problems. If it comes in too fast the friction that occurs when it is travelling through the atmosphere will cause it to burn up. Thus the angle at which it re-enters the atmosphere must not be too steep. On the other hand if the angle is too shallow the spacecraft will 'bounce' back into space.

If the spacecraft re-enters at the correct angle it will slow down and not burn up. Of course it does heat up to some extent, and there is a heat shield at the front of the re-entering craft to protect the space travellers inside (called astronauts by the Americans and cosmonauts by the Russians). The final slowing-down process is achieved by using parachutes.

Travelling in space presents very special problems. Perhaps the most important of

these is navigation—how to be sure of getting to one's destination and back again. The main problem is that the destination is moving. For example, the Moon is travelling around the Earth. Thus a spacecraft travelling to the Moon has to be aimed at a particular point ahead of the Moon's position at take-off.

During the trip the pilot uses special telescopes to find the positions of certain stars in relation to particular places on the Moon. This information can be used by a computer to determine the position of the spacecraft. In practice, the time of launch, the speed and the course of the spacecraft are all worked out before take-off. During the journey only minor course corrections are necessary.

Another problem of space travel is created by the unusual conditions inside the spacecraft. Away from the Earth there is no gravity. Everything is weightless; 'up' and 'down' have no meaning. Astronauts can float to any part of the spacecraft they wish. Thus the controls and instruments can be placed anywhere.

However, the everyday things of life that we take for granted, such as eating, drinking, sleeping and going to the lavatory, become much more difficult. At first all food was eaten from tubes, but recent space trips have shown that it is possible to eat more normal foods in weightless conditions. A space lavatory is constructed so that the waste matter is taken away by suction. An astronaut sleeps in a special sleeping-bag. Once inside the bag he cannot float away while he is asleep.

Above: The shuttle orbiter *Enterprise* separating from its carrier, a specially converted Jumbo jet, during a test flight. The shuttle was designed by U.S. space experts to ferry astronauts to and from an orbiting space station.

Above right: The U.S. space station *Skylab 1*, which was launched in 1973. Astronauts occupied the *Skylab* for a total of 171 days during 1973 and 1974.

Right: A Saturn rocket blasting an Apollo spacecraft into orbit from the Kennedy Space Centre at Cape Canaveral, in Florida.

Far right: Life aboard an Apollo spacecraft was very cramped. Left to right, astronauts Russell L. Schweickart, David R. Scott and James A. McDivitt photographed in the command module.

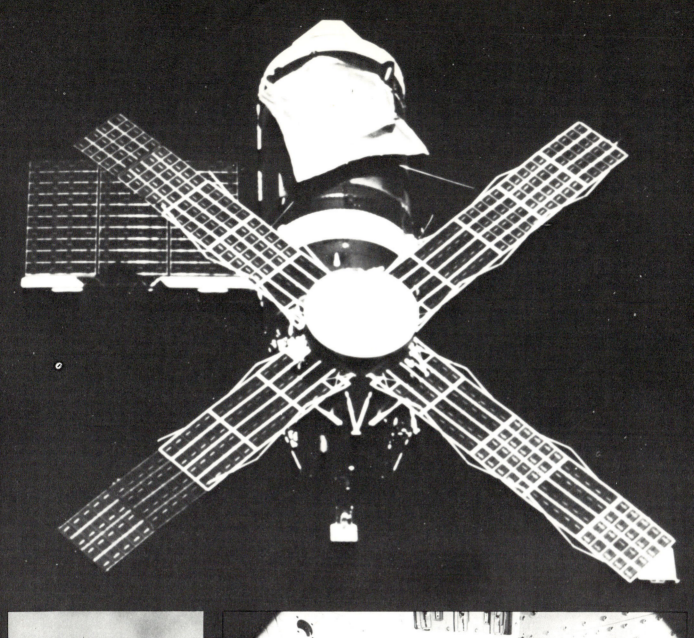

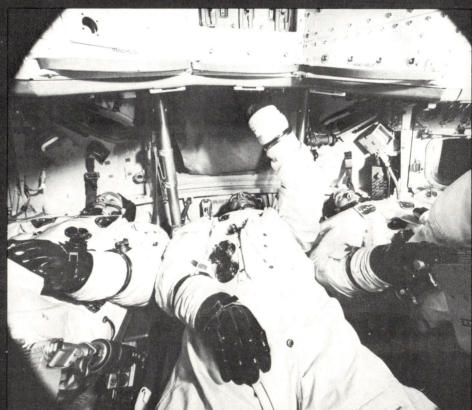

Index

Aborigines 45, 116, 128
Abstract art 109
Acrylic paint 110
Acting 90, 91
Acts of the Apostles 76
Advertising 78
Aeroplanes 182-3
Aeschylus 92
Afghanistan 121
Africa 8, 13, 42-3, 62-3, 69, 126-7, 132, 180
Age of Bronze 109
Agoutis 40
Agriculture 66, 118, 121
Aguardiente 124
Airships 182
Aitken, Prof. Howard 158
Akhenaton, Pharaoh 62
Alaska 13, 122
Albania 130
Alcmene 77
Aldrin, Edwin 'Buzz' 25
Aleppo pines 35
Alexandra the Great 159
Algae 28, 48, 133, 155
Alimentary canal 57
Alkyl benzene 157
Alligators 39
Alpaca 40
Alpha Centauri 20
Alphabet 60
Alpine flowers 37
Alps 118
Altaic language 131
Aluminium 13, 170
Amazon 41, 132
American Indians 116, 122, 124, 125
American War of Independence 136
Amharic 131
Ammonia 17
Amphibians 31
Amphitheatre 92
Anaconda 41
Andes Mts. 40, 41
Angiosperms 28
Anglo-Saxons 66, 67
Animals 26, 30, 31, 38, 51
Annelids 30
Antarctic ocean 10, 33
Antarctica 8, 48, 133, 153
Anteater 40
Antelopes 37, 39, 42, 43
Anthropologists 54
Ants 41
Apennine Mts. 118
Apes 54
Apocrypha 76
Apollo 66
Apollo spacecraft 25
Apostles 77
Arabia 63
Arabs 126, 127, 136, 159
Arachnids 31
Aral Sea 117
Archaeologists 54
Arctic fox 34
Arctic ocean 10, 39, 48
Argentina 124
Argon gas 168
Argyroneta aquatica 50
Aristarchus 24
Arizona 46

Armadillo 40
Armenia 130
Armstrong, Neil 25
Arp, Jean 109
Arsenic 155
Artemis 66
Artesian well 22
Arthropods 30, 31
Arthur, King 67, 77
Arts 82-3
Asia 8, 36, 48, 120-1, 126, 127, 132, 138, 180
Associated Press 79
Assyria 138, 159
Asteroids 16
Astley, Philip 103
Astronauts 25, 186
Athens 76
Atlantic ocean 10, 33
Atmosphere 14, 16, 17
Atom 164-5, 168
Attila 138
Augean Stables 177
Australia 8, 31, 44-5, 128-9, 139, 153, 180
Australopithecus 55
Austria 68, 133, 139

Babbage, Charles 158
Babylonia 62, 138
Bacteria 27, 155, 162
Bactrian camel 46
Badgers 34, 35, 39
Baekeland, Leo 166
Bailey, Nathan 72
Baird, John Logie 72
Bakelite 166
Balance 54
Ballets Russes 102
Balloons, hot-air 182
Bangladesh 120
Banyan tree 37
Barbarians 138
Basques 131
Bassoon 84
Bat 37
Bear 34-8, 48
Beaver 38, 39
Bedouins 127
Bee-eater 35
Beech tree 34
Beginning of the World 109
Belgium 132
Belgrade 133
Bell, Alexander Graham 72, 77
Berbers 126
Bermuda 11
Bible 74, 76
Birch tree 34, 118
Birds 34, 35, 42
Birth of a Nation 101
Bizet, Georges 103
Blackbirds 35
Blood 57
Blue Mts. 44
Blue whale 30
Boa constrictors 41
Bobcats 38
Boers 139
Bolivia 125
Books 74
Bower bird 45
Bracken 35
Brancusi, Constantin 109
Brandenberger, Jacques 166

Brass 170
Brazil 124
Brecht, Bertolt 93
Britain 71, 122, 128, 132, 139
Brittany 130
Brittle-bush 47
Brittle stars 31
Budapest 132
Buddhism 62
Budgerigar 45
Buenos Aires 124
Buffalo 37, 39, 42, 43
Bunyan, Paul 70
Burma 120, 131
Bushmen 126
Buttercup 35
Butterfly 41, 42
Butterfly fish 33
Byzantine artists 112

Cactus 39, 46
Caimans 41
Calcium 13, 18
  carbonate 56
Calculator 160-1
Calder, Alexander 109
Calderón, Pedro 92
Camels 46, 47, 50
Camera 96, 97, 98-9, 103
Camera obscura 103
Camouflage 41, 49
Canada 122
Capercaillies 34
Capybaras 41
Caracas 124
Carbohydrates 57
Carbon dioxide 17, 26
Carbon monoxide 155
Caribbean Sea 10, 38, 48, 49
Carnivores 58
Carobs 35
Caroline Is. 129
Carpathians 118
Carrion eaters 40
Caspian Sea 116, 159
Cassette recordings 89
Cassowary 45, 47
Cattle, wild 37
Caucasus 118, 126, 131
Cedars of Lebanon 35
Cello 84
Cellophane 166
Cells 26
Celluloid 166
Celtic 130
Central America 62, 108, 122, 180
Cerberus 77
Cervantes, Miguel de 92
Cézanne, Paul 113
Chappe, Claude 71
Charcoal 110
Chardonnet, Count Hilaire de 172
Charlemagne 139
Cheetah 43
Chekov, Anton 93
Chiang Kai-shek 139
Chile 125, 132
Chimpanzee 42, 54
China 60-3, 108, 120, 122, 131, 138, 139, 158, 159, 172, 180
Chinchilla 40
Chippewa Indians 70
Chlorine atoms 153
Chlorophyll 26, 51

Chloroplasts 26
Chordates 31
Christianity 63
Chrysalids 49
Cichlids 41
Circus 103
Civilisation 62, 138
Clay 104, 106
Climate 23, 38, 42, 44, 46, 48, 118, 133
Clocks 119
Clouds 11
Coal 146, 147
Cobras 37
Cockatoos 45
Cockerell, Sir Christopher 184
Cockles 30
Coelacanth 50
Coelenterates 30
Colombia 124
Colour blindness 57
Columbus, Christopher 116, 122
Comédie Française 92
Comets 23
Common Market see EEC
Communications 60, 61, 71, 72
Communism 65
Computers 158
Concrete 158
Condor 40
Confucius 62, 63, 64
Conifer 28
Constable, John 113
Constellations 20
Cook, Capt. James 139
Cooke, William 71
Copernicus, Nikolaus 24
Coral 11
Cornwall 67
Cosmic rays 19, 174
Cosmonauts 186
Cotton 172
Coyotes 39
Crane (bird) 49
Crocodile 37, 43
Crustaceans 30, 32
Cubism 113
Cyanide 155
Cyclops 66
Cyclorama 91

Dali, Salvador 113
Danube, river 133
Darwin, Charles 54
Das Kapital 65
Daumier, Honoré 109
David (statue) 109
Dead Sea 43
Debussy, Claude 102
Declaration of Independence 65
Deer 35, 37, 38
Degas, Edgar 109
Democracy 76
Denmark 119, 130, 132, 149
Desalination 153
Descartes, René 64
Deserts 9, 39, 42, 44-7, 121, 128
Detergents 155, 156-7
Diaghilev, Sergei 102
Dictionaries 72
Diesel engine 178
Digestion 56, 57
Disney, Walt 103
Dogs 37, 39, 48
Dolphins 32

Don, river 138
Donatello 109
Donizetti, Gaetano 103
Double bass 84
Drama 92-3
Duck 38, 49
Duck-billed platypus 31
Dust Bowl 117
Dwarf star 18, 20

EEC (European Economic Community) 132
Eagle 34, 35, 37, 38
Eagle ray 33
Early Man 58, 60
Earth 6-18, 24, 186
Earth Mother 62, 66, 69, 70
Earthquakes 11, 12, 13
Earthworm 30
Eastern Roman Empire 112
Ecclesiastes 74
Echinoderm 31
Edison, Thomas A. 88
Eel 33
Egypt 60, 62, 108, 112, 132, 138
Electricity 71, 140, 144, 180
Electromagnetic waves 174
Electron 152, 164
Electronics 186
Elephant 37, 42, 43
Elixir of Love 103
Elk 34
Elm tree 38
Emigration 117
Emu 45, 47
Encyclopedia 77
Energy 18, 19, 26, 148, 149, 158, 174
England 64, 68, 77, 92, 122, 130
Enzymes 57
Epiphytes 40
Equator 13, 127
Erosion 8
Eskimos 70, 122
Esperanto 131
Ethiopia 55, 131
Eucalyptus 44, 47
Euripides 92
Europe 8, 34, 35, 48, 55, 62, 78, 93, 112, 118-9, 132, 138
Evolution 54
Exploration 12
Explorer 1, spacecraft 25
Eyck, Jan van 112

Faraday, Michael 144
Farming 59, 118, 122, 138
Fern tree 44
Fertilisers 154, 155
Feudal system 138
Fiji 129
Film-making 100-101
Fingerprints 57
Finland 34, 130, 180
Fir tree 34, 118
Fire Bird ballet 102
Fish 31, 32, 33, 38, 124
Flamingo 43
Flowers 42
Flute 84
Flying 182-3
Food & Agricultural Organisation 133
Food chain 58, 155
Forests 34-8, 40, 41, 42, 44, 118, 125, 128

Forlanini, Enrico 184
Fossils 6, 22
Fox 35, 38, 40, 48, 49
France 122, 132, 149
French Guiana 124
French Revolution 136
Fresco 112
Frigga 66
Frog 36, 40
Fungi 29, 36
Futurist art 109

Gaea 66
Gaelic 130
Gagarin, Yuri 25
Gainsborough, Thomas 112
Galaxy 6, 20, 21
Galileo Galilei 24, 176
Gall bladder 57
Gamebirds 34
Gamma rays 174
Gannet 34
Gaucho 124
Gaur 37
Gazelle 37
Geese 38, 49
Generating power 144-9
Genes 116
Genesis 76
Gerenuk 43
Germany 61, 78, 93, 132, 138, 139
Geysers 148
Giant Panda 37
Gibbon 54
Giffard, Henri 182
Giraffe 42, 43, 53
Glaciers 9
Globe Theatre 103
Goats 35, 37, 39
God 63, 64
Goddard, Robert 186
Goethe, Johann von 93
Gogh, Vincent van 113
Gold 108, 170
Golden eagle 37
Gopher 39
Gorilla 54
Gospels 76
Gossamer Condor 183
Goths 138
Gouache 110
Granite 9, 109
Grass 9, 29
Grass snake 35
Grasslands 35, 39, 40, 42, 44, 124
Gravity 14, 24
Great Rift Valley 43
Great Train Robbery 100
Great Trek 139
Grebe 43
Greece 60, 62, 64, 76, 77, 92, 112, 130, 138
Greek mythology 66
'Green Revolution' 59
Greenland 122
Grouse 34
Guam Island 10
Guanacos 40
Guitar 84
Gulf of Mexico 39
Gutenberg, Johannes 61, 74
Guyana 124
Gymnosperm 28

Habsburg family 139

Hagfish 32
Haiti 122
Hand-carving 119
Hanuman 69
Hare 49
Harpsichord 102
Hawaii 11, 129
Hawk 38
Hawker Siddeley Harrier 183
Heat 148, 149, 158, 177
Hebrew 63, 76, 131
Helicopters 183
Helium 17, 18
Herb 35
Herbivores 58
Hercules 77
Heredity 116
Hermes 66
Hestia 66
Hiawatha 70
Hibernation 38, 47, 48, 50
Hindenburg airship 182
Hinduism 62, 63, 66
Hippopotamus 42, 43
Hobbes, Thomas 64
Hollywood 101
Holy Grail 77
Holy Roman Empire 138, 139
Homer 77
Hominids 5
Homo sapiens 55, 116
Hooke, Robert 162, 172
Hornbill 36, 42
Horse 53
Hottentot 126
Housing 150
Hovercraft 184, 185
Hubble, Edwin 21
Hungary 131, 133
Hyatt, John Wesley 166
Hydra 77
Hydro-electric dam 159
Hydrochloric acid 57
Hydrofoils 184-5
Hydrogen 17, 18, 152
Hyena 43

Ibex 35
Ibsen, Henrik 93
Iceland 8, 13, 130
Icons 112
Iliad 77
Impressionism 113, 120, 121, 138
India 60, 62, 69, 102
Indian ocean 10
Indonesia 63, 120
Indus River Valley 138
Industrial Revolution 119, 134, 180
Industrial waste 155
Infra-red rays 20
Insects 31, 133
Interlingua 131
Internal combustion engine 178-9
International Date Line 129
Iran 63
Iranians 121
Ireland 122, 132, 180
Iron 13, 18
Irrigation 117
Ishtar 62
Isis 62
Islam 63
Islands 11
Italy 92, 131, 132, 139

Izanami 69

Jaçana 43
Jackal 37
Jaguar 40
Jainism 62
Janssen, Zacharias 162
Jansz, Willem 139
Japan 59, 60, 62, 63, 68, 69, 120
Jazz 84
Jazz Singer 101
Jellyfish 30, 32
Jerboa 35, 47
Jesus of Nazareth 63, 76, 77
Jews 76
Johnson, Samuel 72
Judaism 63
Juniper 35
Jupiter (god) 62, 66
Jupiter (planet) 16, 17, 24

Kalahari Desert 126
Kandinsky, Vassily 113
Kangaroo 44, 45
Kenya 127
Kenyapithecus 55
Kiang 37
Kibbutzim 121
King Kong 100
Kingfisher 43
Kit-fox 39
Klee, Paul 113
Koala 31, 44, 45
Korea 61, 120, 133
Kublai Khan 159

Lakshmana 69
Laminaria 32
Language 71, 130, 131
Lao Tzu 63
Lapland 34
Larch 34
Laser 168-9
Latin America 132
Lead 155, 170
Leech 36
Leeuwenhoek, Anton van 162
Lemming 34, 49
Lemurs 36, 54
Lenin, Vladimir Ilyich 65
Lent 103
Leonardo da Vinci 112
Leopard 36, 37, 43
Liana 41
Lice 31
Lichen 48, 133
Light 18, 23, 174, 176
Lime 34
Linen 172
Lion 37-40, 43
Lippershey, Hans 176
Lithography 74
Little Dancer 109
Liverworts 28
'Living fossils' 50
Lizard 31, 35, 36, 37, 40, 41, 45
Llamas 40
Locke, John 64, 65
London 103, 135, 155
Longfellow, Henry Wadsworth 69, 70
Los Angeles 155
Lugworms 30
Lumière, Louis & Auguste 100
Luxembourg 132

Lymphatic cells 57
Lynx 35
Lyre 44, 45

MacCready, Paul 183
Mach number 183
Madagascar 50, 69
Magnesium 13
Magnetic recording 88
Mahavira 62
Maiman, Theodore 168
Malaysia 120
Malpighi, Marcello 162
Mammals 31, 32, 52
Man 52-6, 64
Mangroves 39
Mao Tse-Tung 138
Maori 129
Marconi, Guglielmo 71
Mariana Is. 129
Mariana Trench 10
Markhor 37
Mars (god) 66
Mars (planet) 16, 24, 25
Marsupials 31, 38, 44
Marx, Karl 65
Masai tribe 127
Maui 70
Mecca 63
Mediterranean Sea 10, 35, 127, 132, 154,
  156
Menelaus, King of Sparta 76
Menstruation 53
Mercury (element) 170, 177
Mercury (god) 66
Mercury (planet) 16, 24
Merlin 67
Mesopotamia 60, 138
Mesquite 47
Mestizo 122, 124
Metals 170
Meteorites 7
Mexico 59, 122
Mice 47
Michelangelo 109, 112
Microscope 162-3
Middle Ages 92, 112, 138
Migration 36, 48, 49
Milky Way (Galaxy) 6, 20, 21
Minerva 66
Mississippi river 132, 133
Mistletoe 35
Mohawk Indians 70
Mole-rats 35
Molière 92
Molluscs 32
Monkeys 37, 41, 42, 54
Monotremes 31, 44
Monteverdi, Claudio 102
Montgolfier, Joseph & Jacques 182
Moon 6, 11, 14, 24, 25, 70, 186
Moore, Henry 109
Morality plays 92
Morse, Samuel 72
Mosquito 31, 49
Mosses 28, 48, 133, 146
Mountains 8, 35, 39
Mozambique 69
Muhammad 63
Music 66, 84-9
Musk ox 38, 48
Muslims 63
Mussels 30
Mutation 55

Myths and legends 66-70

Nanak 63
Natal 139
Natural selection 54, 116
Nectar 29, 41
Negro 122, 124, 126, 129
Neon gas 168
Neptune (god) 66
Neptune (planet) 16, 17
Netherlands 119, 132
Neutron 165
New Guinea 31, 44, 120, 129
New Hebrides 129
New Mexico 46
New Testament 76
New Zealand 129, 139, 148, 149, 153
Newspaper production 78-9
Newts 31
Nickel 13
Niepce, Joseph 103
Nijinsky, Vaslav 102
Nile river 132
Nitrates 155
North America 8, 38, 39, 48, 78, 79, 122-3, 180
North Korea 120
North Pole 48
Northern Lights 23
Norway 34, 130
Novae 20
Nuclear fission 165
Nucleus 21, 26
Nylon 167, 173

Oak tree 34, 35, 38
Oboe 84
Oceans 10, 11, 32
Octopus 30, 32
Odin 66
Oil and natural gas 121, 142, 155
Oil paint 110, 112
Okapi 42
Oklahoma! 93
Old Testament 76
Olympus, Mount 66, 77
Opera 102
Orang-utan 36, 54
Orange Free State 139
Oratorio 103
Orfeo 102
Organ 84
Orinoco river 124
Ostrogoths 138
Owl 35, 38, 46, 49
Oxygen 7, 13, 26, 30, 152, 154, 155

PTFE 167
PVC 167
Pacific ocean 10, 129
Pakistan 120, 138
Palestine 63, 121
Pampas 40, 124
Pancreas 57
Pangolin 36
Paper 60
Paraguay 124, 125
Parchment 60
Paris 76, 100
Parrot 42
Patagonia 40
Peacock 37
Persia (Iran) 63
Petals 29

Petrol engine 178
Petroleum 159
Petrouchka 102
Pharaoh 62
Philippines 59, 120
Philosophy 64
Phoenicians 156
Phosphates 155
Photo-chemical oxidants 155
Photography 72, 94-104
Photosynthesis 26
Phytoplankton 32
Piano 84, 102
Picasso, Pablo 113
Pine marten 34
Piranha 41
Pitchi-pitchi 47
Planet 6, 16-17, 18
Plankton 32, 38
Plant 9, 26, 28, 32, 58
Plant breeders 59
Plastics 166-7
Platypus, duck-billed 31
Pliny the Elder 77
Pluto (god) 66
Pluto (planet) 16, 17
Pointillism 113
Poland 122
Polar bear 38, 48, 51
Polar region 48-9
Polecat 34
Politics 65
Pollen 29
Pollution 154, 155
Polo, Marco 159
Polyester 167
Polynesia 70, 129
Population 124, 133
Poquelin, Jean (Molière) 92
Porcupine 38, 39
Porgy and Bess 93
Porpoise 32
Portugal 131, 180
Poseidon 66
Possum 44
Potassium 13
Pottery 104
Prairie dog 39
Predators 41
Primates 54
Printing 60, 61
Prophets 76
Proteins 57
Protestant Church 63
Protozoa 30, 155
Publishing 74
Puma 38, 39, 40
Punjab 63
Pygmy 116, 120, 126, 129
Pyrenees 118, 131
Python 37

Quasars 21
Quicksilver 176

R101 airship 182
Rabbit 35
Race 116, 117, 120
Racine, Jean 92
Radiant energy 174
Radiation 7
Radio and television 80-1
Radio signals 19, 71
Radio telescope 25

Ragas 102
Rail travel 180-1
Rain 11
Rainbow 176
Rama 69
Ramayana 69
Raphael 112
Rat 47
Rattlesnake 39
Ravana 69
Rayon 173
Recorder 84
'Red Centre' 44
Red Sea 43
Red shift 21
Redwood tree (sequoia) 39
Reformation 134
Reincarnation 62
Reindeer 34, 48, 49
Reindeer moss 38
Religion 62-3, 66, 124
Religious plays 92
Rembrandt van Rijn 112
Renaissance 109, 112
Reporters 78
Reptiles 31
Respiration 26, 58
Resurrection plant 47
Reuters , news agency 79
Revelations 76
Rhea 40, 47
Rhinoceros 36
Rhododendron 37
Rio de Janeiro 124
Rivers 37, 133
Road travel 178-9
Robbia, Luca della 109
Robin Hood 67
Rock 8, 9
Rockets 186
Rodents 40, 41
Rodin, Auguste 109
Roman Catholic Church 63
Romania 109, 131
Romanticism 113
Rome 62, 92, 103, 109, 138, 156
Round Table 68
Rubber 176
Rubens, Peter Paul 112
Russia 34, 65, 93, 102, 120, 131, 132, 138, 180
Rust 177

SRN 1 hovercraft 184
Sacculus 56
Sage-bushes 35
Sahara Desert 42, 127
St. Louis 133
Saliva 57
Salmon 33, 38
Salt 152
São Paulo 124
Satellites 24, 25, 71
Saturn (god) 66, 67
Saturn (planet) 16, 17, 24
Savannah 37
Saxophone 84
Scandinavia 93, 116
Schiller, Friedrich 93
Science and Technology 64, 140
Scribes 60
Sculpture 106-9
Sea 32, 33, 58, 158
Sea-anemones 30

Sea-urchins 31, 32
Seal 39, 48
Seaslugs 32
Seaweed 28, 32
Semaphore 71
Senefelder, Aloys 74
Sepals 29
Seurat, Georges 113
Sewage treatment 154
Shakespeare, William 92, 103
Shark 31, 32
Sheep 35, 37, 39
Shellfish 32
Shintoism 63
Sial 13
Siddhartha Gautama 62
Sikhism 63
Silicon 13, 160
Silk 60, 172
Silver 170
Sima 13
Singapore 120
Sinkiang 120
Sistine Chapel 112
Sita 69, 102
Skeleton 26
Skunk 38, 40
Sky Father 62, 66, 69
Sloths 41
Snails 30
Snake 31, 35, 36, 37, 41
Snake-birds 37
Snow White and the Seven Dwarfs 103
Soap 156-7
Socrates 64
Sodium 13, 18, 152
Solar cells and panels 148
Solomon Is. 129
Sophocles 92
Sound-recording 84
South Africa 50, 139
South America 8, 13, 40, 41, 62, 108, 124, 132, 180
South-East Asia 129
South Korea 120
South Pole 48, 133
Soviet Union see Russia
Space 7
Space travel and exploration 25, 186
Spain 92, 122, 131, 180
Spider 30, 50
Spore 28
Spruce tree 34
Sputnik I satellite 25
Squid 30, 32
Squirrel 34, 35, 36, 38
Sri Lanka 69
Stage director 90
Star 6, 20
Starfish 31, 32
Stearate ions 156
Steel 170
Steppes 35
Stoat 49
Stockton & Darlington Railway 180
Stone Age 112
Stravinsky, Igor 102
Strawberry trees 35
Strindberg, August 93
String instruments 84
Sturm und Drang 93
Sudan 127, 132
Sugar maples 38
Sulphur dioxide 155

Sumeria 130, 138
Sun 6, 11, 14, 18, 19, 48, 70, 159
Sun worship 62, 108
Sunbird 42
Surface tension 152
Surrealism 113
Surtsey Island 8
Sweden 34, 130
Swift 47
Switzerland 68, 139
Symphony orchestra 84

Tahiti Island 139
Taiwan (Formosa) 120, 139
Taoism 62, 63
Tape recording 88, 89
Tasman, Abel 139
Taste buds 56
Telegraphy 71
Telephone 71, 77
Telescope 24, 25, 176
Television 71
Tell, William 68
Tempera 110
Termite 42
Tern 38
Textiles 172-3
Thailand 131
Theatre 92, 93
Tibet 120, 131
Tonga 129
Torah 76
Tortoise 36, 47
Train 180
Transistor 186
Transport 176, 184-5
Travel see Rail travel;
    Road travel; Transport
Tree 34-42
Tree-dwellers 38
Tribal law 60
Trojan Horse 76
Tropics 33
Tuareg 127
Tundra 34
Tungsten 170
Turkey 131
Turner, William 113
Turtle 32, 37

Uganda 126
Ultra-violet rays 19
United Nations 133
United States 13, 25, 71, 93, 101, 122, 148, 153, 180
Universe 6, 21
Ural Mts. 118
Uranus (god) 66
Uranus (planet) 16, 17, 24
Uruguay 124
Utriculus 54

VTOL aeroplane 183
Vega, Lope De 92
Venezuela 124
Venice 103, 132
Venus (goddess) 66
Venus (planet) 6, 16, 24, 25
Vicuñas 40
Vikings 66, 118
Vine 35
Violin 84, 102
Virus 27
Vishnu 69

Visigoths 138
Volcano 11, 13
Vole 35, 38
Vulture 37, 43

Wagner, Richard 103
Wales 67, 130
Wallace, Alfred Russel 54
Walrus 39, 48
War 134
Warhol, Andy 109
Water 8, 10, 152-5
Watson, Thomas 77
Watt, James 158
Weather 23
Webster, Noah 72
West Indies 122
West Side Story 93
Whale 32
Wheatstone, Charles 71
Wind 9
Windmill 149
Witch doctor 60
Wolf 34-9, 49
Wombat 45
World Health Organisation (WHO) 133
World Wars 137
Worms 32, 36
Wright, Orville & Wilbur 182
Writing 71, 76, 138

X-ray 19

Yak 37
Yellow river 139

Zebra 43

Picture Credits
Amora 23, Heather Angel 28, 29, 30, 31, 32, 33, 34, 35, 37, 39, 50, 51, 55, Australian Information Service 84, 86, 90, 102, 104, 131, 182, Australian News and Information Bureau 44, 45, 153, Australian Opera Company 102, Autumn Publishing 11, 74, 75, 82, 91, 94, 95, 96, 97, 132, 167, 174, 176, 177, Clive Barda 88, 89, Barnabys Picture Library 15, 16, 19, 187, Theo Bergstrom 169, Bob Bray 83, 85, 86, 90, 105, 116, 117, British Airways 183, British Antarctic Survey 133, British Broadcasting Corporation 80, 81, British Hovercraft Corporation 185, British Museum 70, 77, 108, 112, 134, 138, British Petroleum Chemicals Ltd. 166, British Petroleum Co. Ltd. 141, 142 167, British Steel Corporation 170, 171, BOC 168, British Tourist Authority 119, 182, Comissariat General de Tourisme 118, Compix 127, Gene Cox 163, Courtaulds 172, Douglas Dickens 2, 3, 60, 61, 62, 63, 173, Dickens Museum 74, Documentation Francais 65, E. Dominguez 114, Anne-Marie Ehrlich 64, 66, 98, 99, 109, 114, 136, 158, 175, Elm Tree Books 87, E.M.I. 100, 101, John Freeman & Co. 71, Autumn Publishing/Hugh Graham 140, 148, 160, 161, German Embassy 87, Gulbenkian Museum, Durham 108, Hale Observatories 18, 20, 21, 159, High Commission for New Zealand 9, 144, 149, Michael Holford Library 18, IBM 158, Italian Institute 114, Keystone 155, K.H. Kurz 36, Mansell Collection 64, 68, 69, 71, 96, 103, 130, Mexican National Tourist Council 67, Moroccan Tourist Office 126, Mirrorpic 79, Tony Morrison 40, 41, 46, 47, 124, 125, Museum of London 135, National Aeronautics & Space Administration, Washington D.C. 7, 14, 16, 17, 24, 25, 186, 187, National Coal Board 146, National Gallery 111, National Geographical Society 72, National Portrait Gallery 139, National Tourist Organisation, Greece 118, Novosti Press Agency 10, 34, 35, 48, 49, 65, 102, 115, Hewlett Packard 160, Autumn Publishing/Pickering 74, 75, 91, 94, 95, 96, 97, 106, 110, 152, 156, 157, Postmaster General 72, 73, Promoter Newspapers Ltd., Bognor 74, 75, P & O 184, Royal College of Music 87, Science Museum 172, 173, 183, Brian Simpson 184, South African Tourist Corporation 42, 43, 126, 127, Southern Electricity Generating Board 145, Swiss National Tourist Office 70, 119, Syndication International 78, 79, Tate Gallery 109, 114, 115, United Kingdom Atomic Energy Authority 164, 165, Union Pacific Railroad 181, United Nations 130, United Nations Picture Library 154, 175, United Press International (U.K.) Ltd. 137, United States Dept. of Agriculture 58, 59, 173, United States Travel Service 8, 122, 123, Victoria & Albert Museum 113, James Webb 26, 27, ZEFA 33, 34, 38, 39, 42, 46.